ANNALS *of* THE NEW YORK ACADEMY OF SCIENCES

T0188426

EDITOR-IN-CHIEF
Douglas Braaten

ASSOCIATE EDITOR
Rebecca E. Cooney

PROJECT MANAGER
Steven E. Bohall

Artwork and design by Ash Ayman Shairzay

The New York Academy of Sciences
7 World Trade Center
250 Greenwich Street, 40th Floor
New York, NY 10007-2157

annals@nyas.org
www.nyas.org/annals

**The New York
Academy of Sciences**

Published by Blackwell Publishing
On behalf of the New York Academy of Sciences

Boston, Massachusetts
2012

ANNALS *of* THE NEW YORK ACADEMY OF SCIENCES

VOLUME
1273

ISSUE
Advances Against Aspergillosis II
Basic Science

ISSUE EDITORS
Karl V. Clemons,[a] Malcolm Richardson,[b] and David S. Perlin[c]
[a]California Institute for Medical Research and Stanford University, [b]University of Manchester, and [c]UMDNJ–New Jersey Medical School

TABLE OF CONTENTS

Academy Membership: Connecting you to the nexus of scientific innovation

Since 1817, the Academy has carried out its mission to bring together extraordinary people working at the frontiers of discovery. Members gain recognition by joining a thriving community of over 25,000 scientists. Academy members also access unique member benefits.

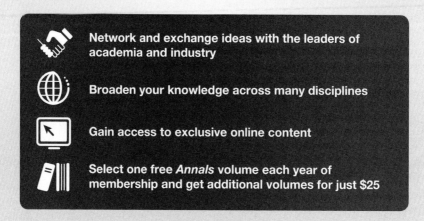

Network and exchange ideas with the leaders of academia and industry

Broaden your knowledge across many disciplines

Gain access to exclusive online content

Select one free *Annals* volume each year of membership and get additional volumes for just $25

Join or renew today at **www.nyas.org**.
Or by phone at **800.843.6927** (**212.298.8640** if outside the US).

Ann. N.Y. Acad. Sci. ISSN 0077-8923

Preface for *Advances Against Aspergillosis*

The fungal genus *Aspergillus* comprises hundreds of species and is an extremely complex taxonomic collection of organisms, including some cryptic species. These molds are ubiquitous in nature, found on decaying organic matter, in the soil, and in marine environments. The saprophytic aspect of *Aspergillus* is very important with respect to environmental ecology. In relationship to humans, *Aspergillus* is important for agricultural, industrial, and medical reasons. In agricultural settings, contamination of food crops with aflatoxin, produced by toxigenic *Aspergillus flavus* is a worldwide problem, which results in the subsequent increased exposure of humans consuming these foods to aflatoxin (a carcinogen). An industrial application of certain *Aspergillus* species is their use in performing specific fermentations, such as in the manufacturing of sake. Medically, about 20 species of *Aspergillus* have been shown to cause disease, with a single species, *A. fumigatus*, responsible for the vast majority of infections and allergies. Although usually an opportunistic pathogen in humans, *A. fumigatus* can be a primary pathogen in animals. Invasive pulmonary aspergillosis is particularly important as a cause of a rapidly fatal disease in avian populations like chickens and turkeys. For humans, aspergillosis has become the fourth leading opportunistic infection of immunocompromised individuals. It is a primarily invasive pulmonary disease, but about one-third of patients have multiple organ involvement, with CNS disease being frequent and often fatal. In spite of advances in antifungal therapy, morbidity and mortality remain high. In addition, *Aspergillus* species cause other diseases in humans that include keratitis, allergic bronchopulmonary aspergillosis in asthma or cystic fibrosis, otomycosis, and several forms of rhinosinusitis and mycotoxicosis.

Recognizing the growing importance of *Aspergillus*, particularly as a cause of human disease, Drs. David W. Denning, William J. Steinbach, and David A. Stevens organized and chaired the inaugural "Advances Against Aspergillosis" conference held in San Francisco in September 2004. With the success of this first meeting, they organized subsequent conferences that have been held every two years. The primary goals have been to bring together clinicians, nurses, clinical laboratorians, epidemiologists, geneticists, molecular biologists, and basic researchers, along with the leading experts in each of these areas from around the world, as well as the introduction of new topics and presenters, and the establishment of new areas of collaborative research and scientific interaction among the different scientific disciplines represented.

The 5th "Advances Against Aspergillosis" was held in Istanbul, Turkey on January 26–28, 2012, with 375 registered attendees from 39 countries and various educational backgrounds attending. The chairs and the scientific committee developed a program that included 40 invited talks, 6 chosen from submitted abstracts, and 144 posters. Sessions included management and treatment of aspergillosis, pathogenesis and immunology, new disease associations, basic biology, non-*fumigatus Aspergillus* species, genomics and proteomics, diagnostics,

doi: 10.1111/nyas.12008

and novel antifungal agents. Poster presentations encompassed all aspects of the study of *Aspergillus*. The full program and programs from past conferences can be viewed at http://www.advancesagainstaspergillosis.org; additional slides and abstracts can be seen at http://www.aspergillus.org.uk.

The papers included in two volumes of *Annals of the New York Academy of Sciences* provide a cross-section of the information presented at the 5th "Advances Against Aspergillosis." Together with the chairs and the scientific committee, we wish to thank the eleven sponsors whose donations allowed the conference to be held, and for scholarships that were given to 28 deserving applicants; the speakers, poster presenters, and attendees, all of whom made the conference a success; the authors of the papers in the two volumes for their time and effort in writing reviews; and the individuals who gave freely of their time in reviewing the manuscripts. We also thank Dr. Douglas Braaten, editor-in-chief, *Annals,* for agreeing to publish the collection of reviews, and for the assistance that he and his staff provided. We look forward to convening the 6th "Advances Against Aspergillosis" in 2014; up-to-date information will be posted at http://www.advancesagainstaspergillosis.org.

KARL V. CLEMONS
Stanford University, Stanford, California

DAVID S. PERLIN
UMDNJ–New Jersey Medical School, Newark, New Jersey

MALCOLM RICHARDSON
University of Manchester, Manchester, United Kingdom

Ann. N.Y. Acad. Sci. ISSN 0077-8923

Evolution of modular conidiophore development in the aspergilli

Steven D. Harris

Department of Plant Pathology, and Center for Plant Science Innovation, University of Nebraska, Lincoln, Nebraska

Address for correspondence: Steven Harris, Center for Plant Science Innovation of Nebraska, E126 Beadle Center, Lincoln, NE 68588-0660. sharri1@unlnotes.unl.edu

Conidiophores are reproductive structures that enable filamentous fungi to produce and disseminate large numbers of asexual spores. The diversity in conidiophore morphology is sufficiently large to serve as a basis for fungal systematics. *Aspergillus* and *Penicillium* species are members of the family Trichocomaceae that form conidiophores with characteristic architecture. Whereas the *Penicillium* conidiophore appears to be a modified branched hyphal structure, the *Aspergillus* conidiophore is seemingly more complex and includes additional cell types. Here, it is proposed that the "aspergillioid" conidiophore may have evolved from a "penicillioid" ancestor via changes in expression of key regulators of the cell cycle and the GTPase Cdc42. Because the transcriptional regulatory network that controls conidiophore development in *Aspergillus* is well characterized, further study of how this network links to regulators of the cell cycle and Cdc42 should provide fundamental insight into the evolution of developmental morphogenesis in fungi (i.e., fungal evo-devo).

Keywords: conidiophore; *Aspergillus*; *Penicillium*; cell cycle control; Cdc42 GTPase module

Understanding the evolution of developmental and morphological processes constitutes one of the primary goals of the field of "evo-devo" (evolution of development). This field has generated significant insight into the molecular mechanisms that cause morphological variation in both animals and plants. One key concept to emerge is the role that changes in gene regulatory sequences play in producing varied morphologies.[1] The resulting temporal and/or spatial alterations in the expression of key genes that determine cell shape and division patterns often trigger profound differences in developmental outcomes. By comparison to animals and plants, remarkably little is known about the evolution of development in the fungi. This is particularly surprising given the wealth of resources and tools that are available for studies of fungal evo-devo. For example, more complete genome sequences are available for fungi than for either animals or plants, and it is far simpler to manipulate individual fungal genes.[2–4] Moreover, fungi produce a striking array of fruiting structures (e.g., mushrooms) that presumably reflect the operation of diverse developmental programs. One central question is whether change in the expression of key morphogenetic genes underlies fungal development to the same extent as it does in animals and plants.

The Trichocomaceae are a family of fungi found within the order Eurotiales.[5] This family contains the well-studied genera *Aspergillus* and *Penicillium* (note that a recent study has proposed that the Trichocomaceae be split into three families: Aspergillaceae, Trichocomaceae, and Thermoascaceae[6]). These genera develop asexual reproductive structures (i.e., conidiophores) that are sufficiently unique to be taxonomically informative. Notably, conidiophore architecture becomes more complex when the anamorphic genus *Paecilomyces* is compared to *Penicillium* and *Aspergillus* (Fig. 1). In *Paecilomyces*, conidiophores are essentially aerial septate hyphae that are branched, with each branch terminating in a phialide that generates spores in a basopetal pattern. *Penicillium* conidiophores consist of septate aerial hyphae that are branched, but are more complex in that each branch terminates in a whorl of phialides. *Aspergillus* conidiophores

doi: 10.1111/j.1749-6632.2012.06760.x

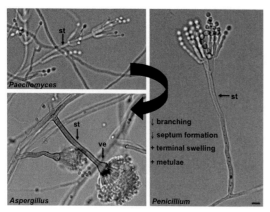

Figure 1. Conidiophore morphology in the *Trichocomaceae*. Examples of conidiophores produced by *Paecilomyces*, *Penicillium*, and *Aspergillus*. The *Paecilomyces* and *Penicillium* isolates have not been identified to the species level. The *Aspergillus* isolate is *A. nidulans* strain FGSC A4. The black arrow denotes the increased complexity of conidiophore morphology. The morphological alterations associated with evolution of the "aspergillioid" conidiophore morphology are indicated (st = stalk, ve = terminal vesicle). Conidiophores were imaged using the "sandwich slide" technique.[25] Bar, 3 μm.

feature several additional modifications that contribute to their increased complexity, including (i) the suppression of branching and septum formation in aerial hyphae (i.e., the stalk), (ii) the formation of a terminal swelling on the stalk that is known as the vesicle, and (iii) in some species (i.e., those that are biseriate), the introduction of an additional tier of cells known as metulae, which in turn bud to generate phialides.

Among the Trichocomaceae, the model fungus *A. nidulans* is widely recognized for its ease of genetic and postgenomic manipulation.[7,8] As a result, considerable progress has been made toward understanding the regulatory network that controls conidiophore development in this fungus.[9–11] Chemical signals (e.g., FluG, Psi factors) and light are two key environmental stimuli that govern the transition from hyphal growth into conidiophore development.[12,13] A network of regulatory factors known as upstream developmental activators (UDAs) is expressed during this transition (Fig. 2). Besides acting to terminate hyphal growth and activate development, these factors (i.e., FlbB, FlbE, FlbD, and FlbC) also collectively promote the expression of the central developmental pathway (CDP; Fig. 2). The pivotal component of the CDP is the transcription factor BrlA, which activates the transcription factors

AbaA and WetA in a pathway that includes multiple feedback loops and is regulated by additional modifiers (i.e., MedA and StuA). Notably, BrlA drives the formation of all cell types up to and including phialides, whereupon AbaA is needed for proper phialide function.[14] Most UDAs and components of the CDP are conserved in other *Aspergillus* spp., as well as in *Penicillium* spp., which implies that the overall architecture of this regulatory network has been conserved within the Trichocomaceae.[10,15] However, the downstream effectors of the UDAs and the CDP are not well characterized. In *A. nidulans*, the expression of genes involved in secondary metabolism and spore maturation appears to be directly regulated by BrlA and/or AbaA.[16,17] Nevertheless, although conidiophore development necessitates dramatic changes in the morphogenetic and cell cycle programs that normally operate in growing hyphae, it is not clear how the UDAs→CDP regulatory network triggers these changes.

A. nidulans has been used as a model fungus for understanding fundamental features of the cell cycle and hyphal morphogenesis.[18,19] These studies have exploited a wealth of available mutants defective in diverse aspects of cell-cycle progression or morphogenesis. The characterization of developmental defects in these mutants has yielded some insight into the identities of the factors that could potentially underlie the evolution of conidiophore morphology in the Trichocomaceae. For example, a mutation affecting the *A. nidulans* homolog of the tyrosine kinase Wee1 (i.e., AnkA) causes dramatic defects in conidiophore architecture (Fig. 3). These include the appearance of septa in the stalk, the pronounced absence of a terminal vesicle, and the failure to properly undergo the transition from "hyphal-like" growth of the stalk and vesicle to "yeast-like" growth of metulae and phialdes. Wee1 is known to phosphorylate the tyrosine-15 (Y15) residues of the

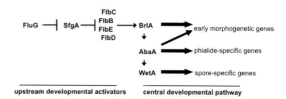

Figure 2. Regulatory network that controls conidiophore development in *A. nidulans*. Note that FlbB and FlbE function as partners to regulate BrlA. See the text for further details.

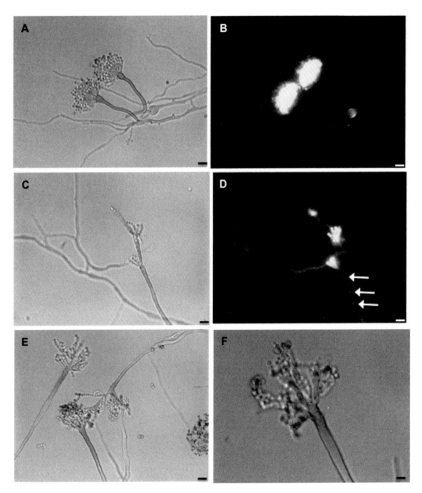

Figure 3. Loss of AnkA function alters conidiophore morphology in *A. nidulans*. (A, B) Wild-type (strain FGSC A28) conidiophores. (C–F) Representative examples of *ankA* mutant (strain APK35) conidiophores. The *ankA* allele in APK35 was originally named *sntA1*.[26] Samples in panels B and D are stained with calcofluor white. Arrows in panel D denote septa present in the conidiophore stalk. Bars, 5 μm, except 2 μm in panel F.

cyclin-dependent kinase CDK1 (i.e., NimX in *A. nidulans*), and in doing so maintains CDK1 in an inactive state during the G2 phase of the cell cycle (Fig. 4).[20] The removal of this phosphorylation by the phosphatase Cdc25 (i.e., NimT in *A. nidulans*) is a requirement for mitotic entry.[20] Accordingly, the conidiophore defects caused by a mutation in AnkA (or by mutation of the Y15 residue of NimX[21]) imply that imposition of a G2 cell cycle delay is needed to inhibit septum formation in the stalk and permit formation of the terminal vesicle. Additional studies show that the ability to precisely regulate the small GTPase Cdc42 is required for normal conidiophore development. RgaA is the *A. nidulans* homolog of the yeast GTPase-activating protein

(GAP) Rga1, which coverts Cdc42 from its active GTP-bound state to an inactive GDP-bound state (Fig. 4).[22] Thus, the absence of Rga1 results in Cdc42 hyperactivity and defective morphogenesis. In *A. nidulans*, *rgaA* deletion mutants generate abnormal conidiophores that do not possess a terminal vesicle and form metulae that resemble hyphae (S. Harris, unpublished results). Moreover, the formation of lateral branches is no longer suppressed in these mutants. Genetic analyses demonstrate that these defects likely result from improperly regulated Cdc42 activity. These observations suggest that Cdc42 activity must be downregulated at two distinct points during conidiophore development: (i) at the tip of stalk to stop polarized growth and

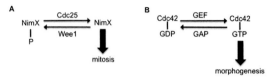

Figure 4. Regulatory modules implicated in the control of morphogenesis during conidiophore development. (A) The CDK module. Wee1 maintains CDK in an inactive phosphorylated state. (B) The Cdc42 GTPase module. Rga1 inactivates Cdc42 by promoting its intrinsic GTPase activity. See the text for further details.

enable formation of the terminal vesicle, and (ii) following the emergence of metulae from the terminal vesicle to facilitate the transition to "budding growth."

Based on the morphological comparisons, regulatory networks, and mutant phenotypes outlined above, our conclusion is that the following hypothesis provides a reasonable framework for understanding the key developmental transition points that could lead to evolution of "aspergillioid" conidiophore from a "penicillioid" ancestor. First, the elimination of septa from the stalk could be achieved by suppressing NimX (CDK1) activity via AnkA. Second, the terminal vesicle could presumably be generated if the tip of the stalk were depolarized by inactivation of the GTPase Cdc42. This could be accomplished by upregulating the GAP RgaA, though downregulation of the Cdc42 GDP-GTP exchange factor Cdc24 might also contribute to inhibition of Cdc42 activity (Fig. 4). Third, the synchronous emergence of buds (i.e., metulae or phialides) from the terminal vesicle could require a burst of Cdc42

activity that must be shut-off (via RgaA) to prevent continued hyphal growth. A morphogenetic checkpoint analogous to that which operates in the yeast *S. cerevisiae* might provide a mechanism to couple the burst of Cdc42 activity with the cell cycle to ensure that each bud possesses only a single nucleus.[23] Finally, the timing of phialide differentiation could determine whether a particular *Aspergillus* species produces uniserate or biserate conidiophores. Because AbaA appears to specify phialide identity,[14] this step could be determined solely through the temporal regulation of *abaA* expression. Each of the transition points proposed in this hypothesis could arise by the evolution of regulatory circuits that drive changes in the temporal and/or spatial expression patterns of genes such as *ankA* or *rgaA*. For example, this could occur by recruiting these genes as targets for regulation by one or more of the transcription factors that compose the UDAs→CDP regulatory network.

The postgenomic era has ushered in an array of sophisticated tools that have significantly influenced the fields of animal and plant evo-devo by facilitating comparative genomic studies. These tools are now being applied in fungi as well, and offer considerable promise for testing hypotheses such as the one outlined above for the evolution of conidiophore development in the aspergilli. For example, the use of transcriptional profiling for comparative expression analyses has yielded important insight into limb development in animals.[24] A similar strategy could be used to compare gene expression during conidiophore development in *A. nidulans* to that of

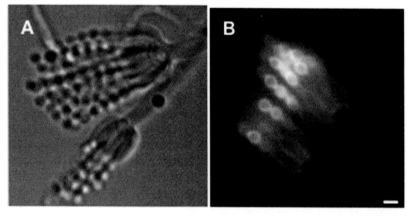

Figure 5. Conidiophores produced by *Hemicarpentales paradoxus* strain NRRL 2162. Samples in panel B are stained with calcofluor white. Bar, 3 μm.

P. chyrsogenum. Such an approach would require the ability to synchronously stage development across both species using specific events such as phialide formation as benchmarks. Another approach would be to compile transcriptional profiles for mutants defective in each component of the UDAs→CDP regulatory network. This would begin to establish regulatory links between individual transcription factors and specific cell cycle or morphogenetic events that underlie conidiophore development. Comparisons of these "regulons" across other *Aspergillus* or *Penicillium* species would then shed light on how these regulatory links might have evolved. A final approach could exploit the availability of unusual species whose conidiophore architecture is not concordant with their phylogenetic placement.[6] As an example, *Hemicarpentales paradoxus* belongs to the *Penicillium* subgenus *Penicillium* (the new name *P. paradoxus* will be proposed[6]), yet produces conidiophores with "aspergillioid" features that include an apparent terminal vesicle (Fig. 5). Assembling the genome sequence of this species would enable studies aimed at determining the molecular basis for this difference. Natural mutations or variation of this type have been extremely informative in studies of animal and plant evo-devo.

Why study the evolution of conidiophore development in the Trichocomaceae or other ascomycetes? One compelling reason is that development culminates in the production of conidiospores, which are the primary means by which many important fungal pathogens of animals and plants spread to the host. Accordingly, any information gleaned from the study of conidiophore development offers the potential to identify novel control strategies that minimize spore production and thereby reduce dissemination. However, a more immediate benefit would be to provide the first clues into how development and morphogenesis have evolved in the fungal kingdom. Although fungi are capable of developing complex multicellular reproductive structures, it is not clear whether the evolution of the underlying developmental pathways follows the same principles that have emerged from animals and plants.[1] Do regulatory networks consist of sets of genes that share common *cis*-acting regulatory sequences? Have regulatory networks been rewired by gains or losses of these regulatory sequences? Are there significant changes in the amino acid sequences of key regulators that affect how pathways have evolved? The answers to these questions will hopefully emerge in the coming years as fungal researchers begin to view development and morphogenesis from an evolutionary perspective.

Acknowledgments

This work was supported by National Science Foundation Grant IOS-0920504.

Conflicts of interest

The author declares no conflicts of interest.

References

1. Carroll, S.B. 2008. Evo-devo and the expanding evolutionary synthesis: a genetic theory of morphological evolution. *Cell* **134:** 25–36.
2. Martin, F., D. Cullen, D. Hibbett, *et al.* 2011. Sequencing the fungal tree of life. *New Phytol.* **190:** 818–821.
3. Nayak, T., E. Szewczyk, C.E. Oakley, *et al.* 2006. A versatile and efficient gene-targeting system for *Aspergillus nidulans.* *Genetics* **172:** 1557–1566.
4. Ninomiya, Y., K. Suzuki, C. Ishii & H. Inoue. 2004. Highly efficient gene replacements in *Neurospora* strains deficient for non-homologous end-joining. *Proc. Natl. Acad. Sci. USA* **101:** 12248–12253.
5. Berbee, M.L., A. Yoshimua, J. Sugiyama & J.W. Taylor. 1995. Is *Penicillium* monophyletic? An evaluation of phylogeny in the family *Trichocomaceae* from 18S, 5.8S and ITS ribosomal DNA sequence sets. *Mycologia* **87:** 210–222.
6. Hourbraken, J. & R.A. Samson. 2011. Phylogeny of *Penicillium* and the segregation of *Trichocomaceae* into three families. *Stud. Mycol.* **70:** 1–51.
7. Todd, R.B., M.A. Davis & M.J. Hynes. 2007. Genetic manipulation of *Aspergillus nidulans:* meiotic progeny for genetic analysis and strain construction. *Nat. Protoc.* **2:** 811–821.
8. Osmani, A.H., B.R. Oakley & S.A. Osmani. 2006. Identification and analysis of essential *Aspergillus nidulans* genes using the heterokaryon rescue technique. *Nat. Protoc.* **1:** 2517–2526.
9. Timberlake, W.E. 1990. Molecular genetics of *Aspergillus* development. *Ann. Rev. Genet.* **24:** 5–36.
10. Yu, J.H., J.H. Mah & J.A. Seo. 2006. Growth and developmental control in the model and pathogenic aspergilli. *Eukaryot. Cell* **5:** 1577–1584.
11. Etxebeste, O., A. Garzia, E.A. Espeso & U. Ugalde. 2010. *Aspergillus nidulans* asexual development: making the most of cellular modules. *Trends Microbiol.* **18:** 569–576.
12. Rodriguez-Urra, A.B., C. Jimenez, M.I. Nieto, *et al.* 2012. Signaling the induction of sporulation involves the interaction of two secondary metabolites in *Aspergillus nidulans.* *ACS Chem. Biol.* **7:** 599–606.
13. Bayram, O., G.H. Braus, R. Fischer & J. Rodriguez-Romero. Spotlight on *Aspergillus nidulans* photosensory systems. *Fungal Genet. Biol.* **47:** 900–908.
14. Sewall, T.C., C.W. Mims & W.E. Timberlake. 1990. *abaA* controls phialide differentiation in *Aspergillus nidulans.* *Plant Cell* **2:** 731–739.

15. Borneman, A.R., M.J. Hynes & A. Andrianopoulos. 2000. The *abaA* homologue of *Penicillium marneffei* participates in two developmental programmes: conidiation and dimorphic growth. *Mol. Microbiol.* **38:** 1034–1047.

16. Stringer, M.A., R.A. Dean, T.C. Sewall & W.E. Timberlake. 1991. Rodletless, a new *Aspergillus* developmental mutant induced by direct gene inactivation. *Genes Dev.* **5:** 1161–1171.

17. Mayorga, M.E. & W.E. Timberlake. 1990. Isolation and molecular characterization of the *Aspergillus nidulans wA* gene. *Genetics* **126:** 73–79.

18. Morris, N.R. & A.P. Enos. 1992. Mitotic gold in a mold: *Aspergillus* genetics and the biology of mitosis. *Trends Genet* **8:** 32–37.

19. Harris, S.D. 2006. Cell polarity in filamentous fungi: shaping the mold. *Int. Rev. Cytol.* **251:** 41–77.

20. Ye, X.S., R.R. Fincher, A. Tang & S.A. Osmani. 1997. The G2/M DNA damage checkpoint inhibits mitosis through Tyr15 phosphorylation of p34cdc2 in *Aspergillus nidulans*. *EMBO J.* **16:** 182–192.

21. Ye, X.S., S.L. Lee, T.D. Wolkow, *et al.* 1999. Interaction between developmental and cell cycle regulators is required for morphogenesis in *Aspergillus nidulans*. *EMBO J.* **18:** 6994–7001.

22. Smith, G.R., S.A. Givan, P. Cullen & G.F. Sprague, Jr. 2002. GTPase-activating proteins for Cdc42. *Eukaryot. Cell* **1:** 469–480.

23. Keaton, M.A. & D.J. Lew. 2006. Eavesdropping on the cytoskeleton: progress and controversy in the yeast morphogenesis checkpoint. *Curr. Opin. Microbiol.* **9:** 540–546.

24. Wang, Z., R.L. Young, H. Xue & G.P. Wagner. 2011. Transcriptomic analysis of avian digits reveals conserved and derived digit identities in birds. *Nature* **477:** 583–586.

25. Lin, X. & M. Momany. 2003. The *Aspergillus nidulans swoC1* mutant shows defects in growth and development. *Genetics* **165:** 543–554.

26. Kraus, P.R. & S.D. Harris. 2001. The *Aspergillus nidulans snt* genes are required for the regulation of septum formation and cell cycle checkpoints. *Genetics* **159:** 557–569.

Ann. N.Y. Acad. Sci. ISSN 0077-8923

ANNALS OF THE NEW YORK ACADEMY OF SCIENCES
Issue: *Advances Against Aspergillosis*

Aspergillus flavus diversity on crops and in the environment can be exploited to reduce aflatoxin exposure and improve health

Hillary L. Mehl,[1] Ramon Jaime,[1] Kenneth A. Callicott,[1,2] Claudia Probst,[1] Nicholas P. Garber,[1] Alejandro Ortega-Beltran,[1] Lisa C. Grubisha,[3] and Peter J. Cotty[1,2]

[1]School of Plant Sciences, The University of Arizona, Tucson, Arizona. [2]Agricultural Research Service, United States Department of Agriculture, Tucson, Arizona. [3]Centre College, Danville, Kentucky

Address for correspondence: Peter J. Cotty, U.S. Department of Agriculture, Agricultural Research Service, School of Plant Sciences, The University of Arizona, P.O. Box 210036, Tucson, AZ 85721. pjcotty@email.arizona.edu

Humans and animals are exposed to aflatoxins, toxic carcinogenic fungal metabolites, through consumption of contaminated food and feed. *Aspergillus flavus*, the primary causal agent of crop aflatoxin contamination, is composed of phenotypically and genotypically diverse vegetative compatibility groups (VCGs). Molecular data suggest that VCGs largely behave as clones with certain VCGs exhibiting niche preference. VCGs vary in aflatoxin-producing ability, ranging from highly aflatoxigenic to atoxigenic. The prevalence of individual VCGs is dictated by competition during growth and reproduction under variable biotic and abiotic conditions. Agronomic practices influence structures and average aflatoxin-producing potentials of *A. flavus* populations and, as a result, incidences and severities of crop contamination. Application of atoxigenic strains has successfully reduced crop aflatoxin contamination across large areas in the United States. This strategy uses components of the endemic diversity to alter structures of *A. flavus* populations and improve safety of food, feed, and the overall environment.

Keywords: *Aspergillus flavus*; aflatoxin; food safety; biocontrol; competitive exclusion

Introduction

The filamentous fungus *Aspergillus flavus* is the primary causal agent of food and feed contamination with the severely toxic fungal metabolites, aflatoxins.[1–4] A wide variety of food crops including maize, cottonseed, peanuts, and tree nuts are susceptible to infection and subsequent aflatoxin contamination.[1] The most common aflatoxin, aflatoxin B$_1$, is a toxic fungal metabolite known to be carcinogenic and teratogenic for both humans and animals.[5,6] It is the only mycotoxin classified as a group 1a human carcinogen by the International Agency for Research on Cancer.[7] Acute health effects of aflatoxin exposure from consumption of highly contaminated food include liver cirrhosis and death.[8] Chronic consumption of sublethal concentrations is associated with liver cancer, growth impairment, and immune suppression.[9–11] In most developed nations, aflatoxins within food and feed are limited by strictly enforced regulations that result in significant economic loss for producers and processors of contaminated crops.[12,13] In developing nations, less effective enforcement of regulations results in chronic exposure to aflatoxins and perennial deleterious effects on human health. The most severe episodes of acute aflatoxicosis with numerous deaths occurred over the past decade.[8,9,14]

Communities of aflatoxin-producing fungi resident in agricultural and native ecosystems are complex assemblages of genotypically and phenotypically diverse individuals.[1] Average aflatoxin-producing potential of *A. flavus* populations is an important determinant of the incidence and severity of aflatoxin contamination events.[3,15,16] In warm regions of the United States, growers of susceptible crops have experienced repeated and severe losses from aflatoxin contamination.[12] For many of these

doi: 10.1111/j.1749-6632.2012.06800.x
Ann. N.Y. Acad. Sci. 1273 (2012) 7–17 © 2012 New York Academy of Sciences.

7

growers, a form of biological control in which native atoxigenic (i.e., do not produce aflatoxins) isolates of *A. flavus* are used to competitively exclude aflatoxin producers is the only viable solution for mitigating contamination.[17–21] Atoxigenic strain applications can be made without increasing the proportion of the crop infected by *A. flavus* and without increasing the overall quantity of *A. flavus* on the crop and in the environment.[22,23] Atoxigenics are naturally associated with crops and, even in the absence of their application, they displace toxigenic *A. flavus* and reduce contamination.[24] Application of native atoxigenic biocontrol strains simply increases the frequency of this natural phenomenon.[17,23] Atoxigenics typically provide over 80% displacement of aflatoxin producers and an associated 80% reduction in contamination with a single application of 10 kg/ha of formulated product (Fig. 1).[18,25,26]

Two atoxigenic strains are currently registered as biopesticides by the EPA,[20] but diverse atoxigenic strains are being developed for biocontrol across the globe. Individuals within *A. flavus* populations vary in aflatoxin-producing potential, virulence, host specialization, competitive ability, and other potentially adaptive traits. Thus, diversity within *A. flavus* populations can be exploited to develop and improve strategies for reducing the average aflatoxin-producing potential of *A. flavus* communities, thereby reducing human exposure to aflatoxins both through consumption of contaminated food and through inhalation of fungal spores and crop fragments.[27–30] This review summarizes current knowledge on diversity and adaptation of *A. flavus* and how human activities, including deliberate modification of *A. flavus* communities as part of aflatoxin management strategies, impact *A. flavus* population structure and the resulting exposure of humans to aflatoxins. Furthermore, future directions for development and improvement of aflatoxin mitigation measures, specifically in the context of biocontrol with atoxigenic strains, are discussed.

Biology of *Aspergillus flavus*

A. flavus is ubiquitous in the environment and proliferates both as a saprophyte[31] and an opportunistic pathogen of plants and animals[32] including humans.[33,34] It is most prevalent in warm regions, especially between latitudes 35 N and 35 S.[1] In addition to influencing human health through exposure

to aflatoxins, *A. flavus* is second only to *A. fumigatus* in nosocomial aspergillosis and is the dominant causal agent among the aspergilli of sinusitis and fungal keratitis.[34] Though primarily studied in agroecosystems, *A. flavus* is common in a variety of natural habitats including the air, soil, and plants of the Sonoran Desert;[22,35] and freshwater, marine,[36] and indoor environments.[34]

The saprophytic phase of the *A. flavus* life cycle occurs primarily in soil where the fungus colonizes organic debris and resides as either mycelia or heavily melanized survival structures called sclerotia.[37,38] When environmental conditions are favorable (e.g., elevated temperature), propagules within debris give rise to conidiophores harboring airborne conidia that are subsequently dispersed throughout the environment.[39] Under appropriate conditions, wind and insect dispersal of conidia to plants results in colonization, infection, and, within susceptible hosts, production of aflatoxins. Conidia produced on plant surfaces serve as inoculum for secondary infections, several cycles of which may occur in a single growing season.[22,40] Infected plant and other organic debris within and on soils serve as reservoirs of *A. flavus* for subsequent dispersal to susceptible hosts and/or nonliving food sources.[41]

The success of genotypically and phenotypically diverse *A. flavus* at each stage in the disease/life-cycle and in different niches affects the structure of aflatoxigenic fungal communities in the environment. Currently, aflatoxin management through application of atoxigenic biocontrol strains aims to displace toxigenic *A. flavus* on a target crop before and during the initial phase of infection.[18] Displacement of toxigenic fungi from a variety of niches rather than just the target crop, and during multiple stages of fungal life cycles including the saprophytic phase, may be necessary to achieve optimal long-term reductions in aflatoxins and other mycotoxins of health and economic concern (e.g., those caused by *Fusarium* species). The genetic and phenotypic diversity of *A. flavus* in various niches provides germplasm of potential value in biocontrol that can be exploited to reduce the prevalence of aflatoxigenic fungi in the environment.

Phenotypic and genotypic diversity in *A. flavus*

Phenotypes of *A. flavus* vary widely, and the species can be divided into two major morphotypes known

Figure 1. Competitive exclusion of aflatoxin producers with an atoxigenic biocontrol strain of *A. flavus* is currently the most effective means for growers to reduce aflatoxin contamination in susceptible crops. Contamination of maize (A) and cottonseed (B) greatly reduces market value, but biocontrol has proved to be an effective management strategy in these crops. Large-scale manufacturing (C, D) and commercial application of biocontrol strain AF36 (E) has been achieved for cotton, maize, and pistachio through a partnership between the Arizona Cotton Research and Protection Council (ACRPC) and the Agricultural Research Service of the U.S. Department of Agriculture (USDA-ARS).[18] Grain coated with the fungus is applied to fields before emergence of susceptible crop components (E). The biocontrol product as applied (left) and following fungal growth (right) are shown in panel (D).

as the L and S strains. S strain isolates produce numerous small (average diameter <400 μm) sclerotia and relatively few conidia, while L strain isolates produce fewer large (average diameter >400 μm) sclerotia and relatively large quantities of conidia.[42] Aflatoxin-producing potential varies widely within *A. flavus* ranging from no aflatoxin production (atoxigenic) to production of over 10⁶ ppb in susceptible crop components.[1,29,43] L strain isolates of *A. flavus* produce on average lower levels of aflatoxins than S strain isolates[3,42] and atoxigenic L strain isolates have been reported from many regions.[4,15,44,45] In contrast, S strain isolates of *A. flavus* produce relatively high levels of aflatoxins, and reports of atoxigenic S strain isolates are lacking.[42] There are other species of aflatoxin-producing fungi with S strain morphology, and these also produce large quantities of aflatoxins.[16] Aflatoxin-producing potential is not associated with either virulence[30,42] or competitive ability during crop infection.[30] The genetic basis for atoxigenicity varies among isolates and may include both single polymorphisms that disrupt production of key enzymes[46] and large deletions in the aflatoxin biosynthesis gene cluster.[47–49]

Determining the primary etiologic agent(s) (i.e., causal agents) of aflatoxin contamination is critical for predicting risk of contamination events and for designing and implementing management strategies. However, the most prevalent type of *A. flavus* infecting a crop is not necessarily the most important etiologic agent of contamination. For example, even when the S strain causes a small proportion (5 or 10%) of the total *A. flavus* infections, it can be the most important causative agent of a contamination event since the S strain produces very high aflatoxin concentrations.[3,14,16,24,50] Thus, aflatoxin management strategies that reduce frequencies of the S strain morphotype may be particularly effective at reducing contamination.[51]

Great genetic diversity among *A. flavus* isolates can be resolved from molecular characteristics including chromosomal karyotypes,[52] mitochondrial, and nuclear restriction fragment length polymorphisms (RFLPs),[53–56] microsatellites,[57,58] nucleotide sequence data,[59,60] and the presence or absence of particular genes and/or indels.[16,47,61] A vegetative incompatibility system mediates self/nonself recognition among genotypically diverse *A. flavus* individuals, and membership of isolates within vegetative compatibility groups (VCGs) provides additional criteria to assess genetic diversity within *A. flavus* populations. Individuals that undergo anastomosis must possess identical alleles at the loci governing vegetative compatibility in order to form a stable heterokaryon and allow gene flow between those individuals. In contrast, gene flow is restricted between dissimilar individuals.[62,63] Phenotypic characteristics, including sclerotial morphology and ability to produce aflatoxins, are typically conserved within a VCG,[64–67] and thus VCGs are commonly treated as functional genetic/ecological/epidemiological units.

Vegetative compatibility analyses provide insight into population genetic diversity as well as changes in compositions of crop-associated *A. flavus* associated with various events.[54,67–71] *A. flavus* populations are composed of many VCGs, and multiple VCGs may occur within a single crop component or aliquot of soil.[70] Some VCGs are common in the environment whereas others are rarely isolated,[67,70,71] and relative frequencies of morphotypes and VCGs vary among crops, fields, regions, seasons, and years.[15,22,44,67,72–75] Thus, each agroecosystem has its own unique, continuously fluctuating assemblage of genetically diverse *A. flavus* that must be managed to minimize crop contamination. Selection of agroecosystem-adapted atoxigenic strains for biocontrol should therefore include region-wide analyses of the genetic and phenotypic diversity within *A. flavus* populations.

DNA sequence data confirm that isolates within a VCG are closely related and genetically distinct from other VCGs. For example, with sequence data from three loci, 36 L strain isolates from six VCGs were divided into four lineages.[59] RFLPs provide greater resolution and were able to separate 75 isolates from 44 VCGs into distinct VCG-defining lineages with sufficient resolution to detect variation among isolates within a VCG.[54] Array comparative genomic hybridization (aCGH) indicated nearly identical gene content within a VCG but up to 2% differences in which genes are present between VCGs.[76] A population study using 24 microsatellite (or simple sequence repeat (SSR)) markers to examine 243 isolates from three VCGs found genetic variation among and within VCGs, but each of the three VCGs were highly resolved, and genetic exchange among them was not detected.[58] Isolates were from sympatric populations, diverse geographic origins, and multiple years, indicating VCGs are genetic lineages that can be widespread in the environment and persist over time. Analysis of variability in microsatellite loci within and among the three VCGs allowed estimates of the time of divergence between the VCGs of 10,000 to 60,000 years before present. Thus, divergence predates the advent of agriculture.[58] Following divergence, VCGs have, at least in some cases, evolved clonally, sometimes in association with agroecosystems. The potential for a sexual stage has been shown in laboratory crosses,[77,78] but the importance of sexual reproduction in shaping *A. flavus* populations in nature and in the process of *A. flavus* evolution and adaptation to agroecosystems is unclear.

Geiser *et al.* raised sexual recombination as a potential stumbling block to the use of atoxigenic isolates as biocontrol agents, presumably due to the creation of highly competitive toxigenic recombinants.[60] In our work, exotic atoxigenic strains have never been introduced to a region; only atoxigenic strains native to target agroecosystems are used. Indeed the most widely distributed atoxigenics are preferred as distribution is used as a proxy for success within the environment. As such, these VCGs have coexisted with the aflatoxins producers they are displacing for over 10,000 years, and no new opportunity for recombination is created by atoxigenic strain use. While sexual reproduction has been reported in the lab under stringent conditions,[77] including between the two biocontrol strains currently approved for use in the United States and other *A. flavus* isolates,[78] genetic data from natural populations provide no evidence for genetic exchange between VCGs.[58] Sexual reproduction, if it occurs in natural populations, is apparently at a low frequency. Even under highly favorable conditions, low fertility[79] and required long time frames[77] observed in laboratory crosses suggest rare recombination. Sexual reproduction may contribute to the evolution of new genotypes within *A. flavus*, but current population genetic studies suggest the importance of sexuality lies on an evolutionary time scale, not an epidemiological one.[58] However, if recombination did occur between an atoxigenic genotype and an aflatoxins producer, the result would be an increased diversity of atoxigenic haplotypes.

Adaptation in *A. flavus*

The broad spectrum of degrading enzymes produced by *A. flavus* isolates as well as their ability to obtain nutrition both pathogenically and saprophytically from a wide variety of hosts and substrates suggests a lack of specialization by the species.[32,80,81] However, variability among isolates in production of pectinases, hydrolytic enzymes involved in maceration of plant host tissues, suggests differential adaptation to plant hosts.[80,82] Ability of isolates to spread between and rot cotton locules is correlated with the production of a specific pectinase P2c,[83,84]

and P2c knock-out mutants have reduced virulence to cotton.[85] Variable pectinase production among *A. flavus* isolates suggests some isolates are more adapted to plants whereas others are more adapted to niches where ability to macerate plant tissues provides less of an advantage.[82]

Phenotypic variation among *A. flavus* isolates provides further evidence for niche differentiation. For example, production of large quantities of sclerotia and aflatoxins may confer an adaptive advantage in soils where long-term survival and defense against insect grazing may be essential for success. In some studies, soil populations of *A. flavus* produced, on average, more aflatoxins and sclerotia than *A. flavus* from crops.[86,87] Production of high levels of aflatoxins and sclerotia and a lack of pectinase P2c production by about 50% of S strain isolates from Arizona suggest at least some members of this morphotype are more adapted to soil than to crop environments.[42,82] Aflatoxin producers with S morphology have been identified as the primary etiologic agents of several contamination events.[14,50] If these isolates are best adapted to soil environments, they may be particularly vulnerable to competitive displacement by atoxigenics in the crop.[4,24] Conversely, in soils during fallow periods, S strain may be favored over crop-adapted atoxigenics, and therefore applied atoxigenic strains may have comparatively less persistence in the environment. The identity and frequency of VCGs differ between soil and crop populations even within a single field, and some VCGs common in soil are not easily detected on the crop and vice versa.[70,87] Such soil resident *A. flavus* genotypes may be overwintering between preferred hosts (either animal or plant) or may be better adapted as saprophytes than as pathogens. Regardless of that, competition between soil-adapted *A. flavus* and applied atoxigenic biocontrol strains during periods of overwintering probably influences the persistence of atoxigenics in the environment. Thus, utilization of both soil- and crop-adapted atoxigenics may be necessary to achieve optimal long-term modification of the aflatoxin-producing fungal community and to both increase additive benefits and reduce the necessity of yearly applications.

Ability of genotypes to compete during host infection may have a greater influence on the *A. flavus* population structure than virulence. This is a type of host specialization in which a genotype has

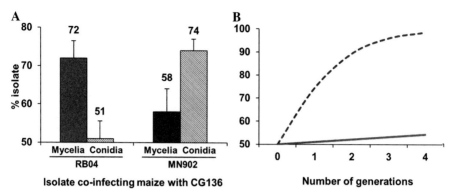

Figure 2. Differential ability of *A. flavus* isolates to compete during maize kernel infection and sporulation (A) and the predicted influence of these competitive differences on *A. flavus* population dynamics over time (B). Isolates RB04 and MN902 were each co-inoculated in equal proportions on maize kernels with an isolate from a common toxigenic VCG, CG136. Isolate percentages (A) from kernel-infecting mycelia and conidia produced during infection were determined by quantifying isolate-specific single nucleotide polymorphisms from total *A. flavus* mycelia and conidia DNA with pyrosequencing. RB04 outcompeted CG136 during maize kernel infection but not during sporulation, whereas MN902 was significantly more competitive during sporulation than during co-infection of kernels with CG136. The predicted influence of competitive differences detected after one cycle of infection and reproduction (A) on proportions of RB04 (solid line) and MN902 (dashed line) within the *A. flavus* population over time are shown in panel B. Conidia produced during host infection contribute to secondary infections, many of which may occur in a single growing season, and presumably shift population structure with each subsequent cycle of infection and reproduction. Calculations are based on the assumption that the isolates initially encounter the crop in equal proportions (50% each) and that the *A. flavus* "population" comprises only CG136 and either RB04 or MN902. And although RB04 comprised a greater percentage of the total *A. flavus* infecting the crop (relative to CG136) after one cycle of infection and reproduction, the superior ability of MN902 to compete during sporulation may contribute to its success on the crop over time. Data are derived from Ref. 30.

adaptations that confer advantage during colonization and infection of some hosts but not necessarily others.[30,88] Indeed, differential competitive ability among isolates on plant hosts indicates crops can select for certain genotypes within *A. flavus* populations through influences on outcomes of competition (Fig. 2). The notion that hosts influence the *A. flavus* population structure is supported by studies in which both aflatoxin-producing potential of *A. flavus* populations[89,90] and quantities and frequencies of morphological and genetic types[73,74,91] were found to vary among crop hosts. Furthermore, though lineages are not exclusively associated with a particular host, certain genotypes of *A. flavus* are more likely to be associated with specific hosts or habitats than others[36] suggesting divergence in host specialization within *A. flavus* populations.

During competition for plant host substrates, some genotypes of *A. flavus* are highly successful during host invasion and tissue ramification whereas others compete poorly during capture of substrates but are highly competitive during sporulation.[30] These differential strategies in response to competition likely have important impacts on the *A. flavus* population structure and the epidemiology

of *A. flavus* infection and aflatoxin contamination. *A. flavus* individuals highly competitive during capture of host substrates have the greatest influence on aflatoxin content within infected host tissues.[30,42] In contrast, an isolate highly competitive during sporulation may come to dominate the population during multiple cycles of reproduction[22,40] even if it is a relatively poor competitor during crop infection[30] (Fig. 2). However, when environmental conditions are not conducive to sporulation, dispersal, and secondary infection, genotypes dominant within host tissues are more adapted to long-term survival in cropping systems.[41] Rather than one strategy being superior over the other, differential behavior of isolates during competition may reflect niche partitioning that allows for maintenance of multiple genotypes within *A. flavus* populations over time and space.[92] Diverse adaptive strategies to various hosts presumably influence ability of atoxigenics to infect and multiply on crops and to persist in the environment. Competitive success during all stages of the *A. flavus* disease/life cycle and on multiple hosts and nonliving substrates are important criteria for the selection of atoxigenic isolates for potential use in aflatoxin mitigation.

Impact of human activity on *A. flavus* populations

Crop rotations influence aflatoxin contamination by altering both density and structure of *A. flavus* populations. For example, colony-forming units (CFU) of *A. flavus* increased in soils continuously rotated from maize to groundnut,[93] and, likewise, CFU of *A. flavus* were greater in residues from plots cropped continuously to maize than from plots cropped with a soybean–maize rotation.[94] In southern Texas, significantly higher CFU of *A. flavus* occurred in soils previously cropped to maize than those previously cropped to cotton or sorghum.[74,91] Higher population densities of *A. flavus*, however, do not always translate to higher crop contamination. Aflatoxin contamination is influenced by both density and structure of *A. flavus* populations, and crops that favor high S strain incidences increase the aflatoxin-producing potential of the *A. flavus* population.[15] In southern Texas, soils previously cropped to cotton and sorghum had higher frequencies of the S strain

than those cropped to maize (Fig. 3).[74,91] Crop rotations may be manipulated to lower incidences of the highly toxigenic S strain and to favor success of applied atoxigenic strains. This is just one potential benefit that may arise from extended research on the influence of crop rotations on the *A. flavus* population structure.

Climate influences the density and structure of *A. flavus* populations as well as the extent to which fungi produce aflatoxins within crops. Thus, climate change has the potential to alter both the incidence and severity of aflatoxin contamination events.[91,95] Contamination is favored by hot and dry climates.[40,96,97] Hot climates also favor higher densities of *A. flavus* and higher incidences of the S strain[22,50] indicating periods of increased soil temperature drive selection of the highly toxigenic S strain.[91] Although certain crops apparently favor the S strain (Fig. 3), it is difficult to separate crop rotation influences from environmental influences. The S and L strains have different adaptations,[22,24,95]

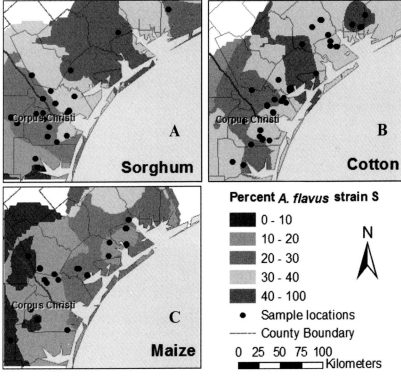

Figure 3. Percentage of the population of *A. flavus* composed of the S strain in soils of southern Texas previously cropped to sorghum (A), cotton (B), and maize (C). Rotations with sorghum and cotton favored increased incidences of the high aflatoxin-producing S strain morphotype. Rotation to corn favored reduced S strain incidence and increased frequences of the L strain morphotype that has lower average aflatoxin-producing potential. Maps of data previously reported in Ref. 74.

with the S strain better adapted to crops grown in warm environments. S strain isolates are most prevalent in warm regions of western and central Arizona and southern Texas where cotton and sorghum are major crops.[15,50,74,75] However, aflatoxin producers with S strain morphology also dominate in portions of East Africa[14] and in northern Texas where maize is an important crop. Repeated and severe contamination in eastern Kenya is due in part to high incidences of isolates with S strain morphology.[3,14] Thus, temperature influences on prevalence of these fungi may be a mechanism through which climate change will threaten food safety and human health worldwide.[91,95]

Conclusions: Future directions for biological control of aflatoxin-producing fungi

There are many genetically diverse atoxigenic VCGs of *A. flavus*, and atoxigenic isolates have been found in every target region examined to date.[4,15,44,49] In many regions, there are sufficient endemic, well-adapted atoxigenic strains to permit treatment with complex strain mixtures and to rotate mixtures between seasons and crops. This strategy has the potential to allow for additive and long-term reductions in the aflatoxin-producing potential of *A. flavus* communities associated with crops and throughout the environment and, in so doing, elimination of frequent human exposure to unsafe aflatoxin concentrations. As described in this review, phenotypic and genotypic variation among *A. flavus* individuals confers differential adaptation to hosts, soils, and climate. These divergences influence abilities of individual atoxigenics to compete across landscapes and through crop rotations. Currently, changes to fungal community structures caused by atoxigenic strain application gradually decline over three years.[17,98] Improving persistence of applied atoxigenic biocontrol strains may be possible through the selection of agroecosystem-adapted strains, use of strain mixtures that include isolates adapted to different niches/hosts or environmental fluctuations within the agroecosystem, and the implementation of agronomic practices that favor atoxigenics and suppress highly toxigenic aspergilli (i.e., the S strain). The future of atoxigenic strain technology should include assessment of phenotypic and genotypic diversity within *A. flavus* populations in order to develop formulations containing

mixtures of atoxigenics with adaptive traits that will allow for long-term residence in target regions, thus increasing protection from aflatoxins at reduced cost.

Conflicts of interest

The authors declare no conflicts of interest.

References

1. Cotty, P.J., P. Bayman, D.S. Egel & K.S. Elias. 1994. Agriculture, aflatoxins and *Aspergillus*. In *The Genus* Aspergillus: *From Taxonomy and Genetics to Industrial Application*. K.A. Powell, A. Renwick & J.F. Peberdy, Eds.: 1–27. Plenum Press. New York.
2. Klich, M.A. 2007. *Aspergillus flavus*: the major producer of aflatoxin. *Mol. Plant Pathol.* **8:** 713–722.
3. Probst, C., F. Schulthess & P.J. Cotty. 2010. Impact of *Aspergillus* section *Flavi* community structure on the development of lethal levels of aflatoxins in Kenyan maize (*Zea mays*). *J. Appl. Microbiol.* **108:** 600–610.
4. Probst, C., R. Bandyopadhyay, L.E. Price & P.J. Cotty. 2011. Identification of atoxigenic *Aspergillus flavus* isolates to reduce aflatoxin contamination of maize in Kenya. *Plant Dis.* **95:** 212–218.
5. McKean, C., L. Tang, M. Tang, M. Billam, Z. Wang, C.W. Theodorakis, R.J. Kendall & J.S. Wang. 2006. Comparative acute and combinative toxicity of aflatoxin B_1 and fumonisin B_1 in animals and human cells. *Food Chem. Toxicol.* **44:** 868–876.
6. Wang, J.-S. & L. Tang. 2004. Epidemiology of aflatoxin exposure and human liver cancer. *Toxin Rev.* **23:** 249–271.
7. International Agency for Research on Cancer. 2002. *Traditional Herbal Medicines, Some Mycotoxins, Naphthalene and Styrene*. International Agency for Research on Cancer. Lyon, France.
8. Centers for Disease Control and Prevention. 2004. Outbreak of aflatoxin poisoning—eastern and central provinces, Kenya, January–July 2004. *Morb. Mortal. Wkly. Rep.* **53:** 790–793.
9. Khlangwiset, P., G.S. Shepard & F. Wu. 2011. Aflatoxins and growth impairment: a review. *Crit. Rev. Toxicol.* **41:** 740–755.
10. Liu, Y. & F. Wu. 2010. Global burden of aflatoxin-induced hepatocellular carcinoma: a risk assessment. *Environ. Health Perspect.* **188:** 818–824.
11. Williams, J.H., T.D. Phillips, P.E. Jolly, J.K. Stiles, C.M. Jolly & D. Aggarwal. 2004. Human aflatoxicosis in developing countries: a review of toxicology, exposure, potential health consequences, and interventions. *Am. J. Clin. Nutr.* **80:** 1106–1122.
12. Robens, J. & K.F. Cardwell. 2005. The costs of mycotoxin management in the United States. In *Aflatoxin and Food Safety*. H.K. Abbas, Ed.: 1–12. CRC Press. Boca Raton.
13. van Egmond, H.P., R.C. Schothorst & M.A. Jonker. 2007. Regulations relating to mycotoxins in food: perspectives in a global and European context. *Anal. Bioanal. Chem.* **389:** 147–157.
14. Probst, C., H. Njapau & P.J. Cotty. 2007. Outbreak of an acute aflatoxicosis in Kenya in 2004: identification

of the causal agent. *Appl. Environ. Microbiol.* **73:** 2762–2764.

15. Cotty, P.J. 1997. Aflatoxin-producing potential of communities of *Aspergillus* section Flavi from cotton producing areas in the United States. *Mycol Res.* **101:** 698–704.

16. Probst, C., K.A. Callicott & P.J. Cotty. 2012. Deadly strains of Kenyan *Aspergillus* are distinct from other aflatoxin producers. *Eur. J. Plant Pathol.* **132:** 419–429.

17. Cotty, P.J. 2006. Biocompetitive exclusion of toxigenic fungi. In *The Mycotoxin Factbook: Food and Feed Topics.* D. Barug, D. Bhatnagar, H.H.P. van Egmond, *et al.*, Eds.: 179–197. Wageningen Academic Publishers. Wageningen.

18. Cotty, P.J., L. Antilla & P.J. Wakelyn. 2007. Competitive exclusion of aflatoxin producers: farmer driven research and development. In *Biological Control: A Global Perspective.* C. Vincent, N. Goettel & G. Lazarovits, Eds.: 241–253. CAB International. Oxfordshire.

19. Dorner, J.W., R.J. Cole, W.J. Connick, *et al.* 2003. Evaluation of biological control formulations to reduce aflatoxin contamination in peanuts. *Biol. Control* **26:** 318–324.

20. Dorner, J.W. 2004. Biological control of aflatoxin contamination of crops. *Toxin Rev.* **23:** 425–450.

21. Yin, Y., L. Yan, J. Jiang & Z. Ma. 2008. Biological control of aflatoxin contamination of crops. *J. Zheijang Univ. Sci. B* **9:** 787–792.

22. Bock, C.H., B. Mackey & P.J. Cotty. 2004. Population dynamics of *Aspergillus flavus* in the air of an intensively cultivated region of south-west Arizona. *Plant Pathol.* **53:** 422–433.

23. Cotty, P.J. 1994. Influence of field application of an atoxigenic strain of *Aspergillus flavus* on the populations of *A. flavus* infecting cotton bolls and on aflatoxin content of cottonseed. *Phytopathology* **84:** 1270–1277.

24. Cotty, P.J., C. Probst & R. Jaime-Garcia. 2008. Etiology and management of aflatoxin contamination. In *Mycotoxins: Detection Methods, Management, Public Health and Agricultural Trade.* J.F. Leslie, R. Bandyopadhyay & A. Visconti, Eds.: 287–299. CAB International. Oxfordshire.

25. Smith, R. 2009. *Non-Toxic Fungus May Hold Key to Aflatoxin Contamination.* Southwest Farm Press, Available at: http://southwestfarmpress.com/grains/non-toxic-fungus-may-hold-key-aflatoxin-contamination.

26. Stalcup, L. 2011. New aflatoxin armor available. *Corn and Soybean Digest*, Available at: http://cornandsoybeandigest.com/crop-chemicals/new-aflatoxin-armor-available.

27. Desai, M.R. & S.K. Gosh. 2003. Occupational exposure to airborne fungi among rice mill workers with special reference to aflatoxin producing *A. flavus* strains. *Ann. Agric. Environ. Med.* **10:** 159–162.

28. Fischer, G. & W. Dott. 2003. Relevance of airborne fungi and their secondary metabolites for environmental, occupational and indoor hygiene. *Arch. Microbiol.* **179:** 75–82.

29. Lee, L.S., J.H. Wall, P.J. Cotty & P. Bayman. 1990. Integration of ELISA with conventional chromatographic procedures for quantification of aflatoxin in individual cotton bolls, seeds, and seed sections. *J. Assoc. Off. Anal. Chem.* **73:** 581–584.

30. Mehl, H.L. & P.J. Cotty. 2010. Variation in competitive ability among isolates of *Aspergillus flavus* from different vegetative

31. Klich, M.A. 2002. Biogeography of *Aspergillus* species in soil and litter. *Mycologia* **94:** 21–27.

32. St. Leger, R.J., L. Joshi & D.W. Roberts. 1997. Adaptation of proteases and carbohydrates of saprophytic, phytopathogenic, and entomopathogenic fungi to the requirements of their ecological niches. *Microbiology* **143:** 1983–1992.

33. Hedayati, M., A.C. Pasquallotto, P.A. Warn, *et al.* 2007. *Aspergillus flavus*: human pathogen, allergen and mycotoxin producer. *Microbiology* **153:** 1677–1692.

34. Sepahvand, A., M. Shams-Ghahfarokhi, A. Allameh, *et al.* 2011. A survey on distribution and toxigenicity of *Aspergillus flavus* from indoor and outdoor hospital environments. *Folia Microbiol.* **56:** 527–534.

35. Boyd, M.L. & P.J. Cotty. 2001. *Aspergillus flavus* and aflatoxin contamination of leguminous trees of the Sonoran Desert in Arizona. *Phytopathology* **91:** 913–919.

36. Ramírez-Camejo, L.A., A. Zuluaga-Montero, M. Lázaro-Escudero, *et al.* 2012. Phylogeography of the cosmopolitan fungus *Aspergillus flavus*: is everything everywhere? *Fungal Biol.* **116:** 452–463.

37. Ashworth, L.J., Jr., J.L. McMeans & C.M. Brown. 1969. Infection of cotton by *Aspergillus flavus*: epidemiology of the disease. *J. Stored Prod. Res.* **5:** 193–202.

38. Horn, B.W. 2003. Ecology and population biology of aflatoxigenic fungi in soil. *Toxin Rev.* **22:** 351–379.

39. Wicklow, D.T. & J.E. Donahue. 1984. Sporogenic germination of sclerotia in *Aspergillus flavus* and *A. parasiticus*. *T. Br. Mycol. Soc.* **82:** 621–624.

40. Diener, U.L., R.J. Cole, T.H. Sanders, *et al.* 1987. Epidemiology of aflatoxin formation by *Aspergillus Flavus*. *Annu. Rev. Phytopathol.* **25:** 249–270.

41. Jaime-Garcia, R. & P.J. Cotty. 2004. *Aspergillus flavus* in soils and corncobs in south Texas: implications for management of aflatoxins in corn-cotton rotations. *Plant Dis.* **88:** 1366–1371.

42. Cotty, P.J. 1989. Virulence and cultural characteristics of two *Aspergillus flavus* strains pathogenic on cotton. *Phytopathology* **79:** 808–814.

43. Probst, C. & P.J. Cotty. 2012. Relationship between *in vivo* and *in vitro* aflatoxin production: reliable prediction of fungal ability to contaminate maize with aflatoxins. *Fungal Biol.* **116:** 503–510.

44. Atehnkeng, J., P.S. Ojiambo, T. Ikotun, *et al.* 2008. Evaluation of atoxigenic isolates of *Aspergillus flavus* as potential biocontrol agents for aflatoxin in maize. *Food Addit. Contam.* **25:** 1266–1273.

45. Cotty, P.J. 1990. Effect of atoxigenic strains of *Aspergillus flavus* on aflatoxin contamination of developing cottonseed. *Plant Dis.* **74:** 233–235.

46. Ehrlich, K.C. & P.J. Cotty. 2004. An isolate of *Aspergillus flavus* used to reduce aflatoxin contamination in cottonseed has a defective polyketide synthase gene. *Appl. Microbiol. Biotechnol.* **65:** 473–478.

47. Chang, P., B.W. Horn & J.W. Dorner. 2005. Sequence breakpoints in the aflatoxin biosynthesis gene cluster and flanking

compatibility groups during maize infection. *Phytopathology* **100**: 150–159.

regions in non-aflatoxigenic *Aspergillus flavus* isolates. *Fungal Genet. Biol.* **42**: 914–923.

48. Donner, M., J. Atehnkeng, R.A. Sikora, *et al.* 2010. Molecular characterization of atoxigenic strains for biological control of aflatoxins in Nigeria. *Food Addit. Contam.* **27**: 576–590.

49. Yin, Y., T. Lou, L. Yan, *et al.* 2009. Molecular characterization of toxigenic and atoxigenic *Aspergillus flavus* isolates collected from peanut fields in China. *J. Appl. Microbiol.* **107**: 1857–1865.

50. Jaime-Garcia, R. & P.J. Cotty. 2006. Spatial distribution of *Aspergillus flavus* and its toxigenic strains on commercial cottonseed from South Texas and its relationship to aflatoxin contamination. *Plant Pathol.* **55**: 358–366.

51. Garber, R.K. & P.J. Cotty. 1997. Formation of sclerotia and aflatoxins in developing cotton bolls infected by the S strain of *Aspergillus flavus* and potential for biocontrol with an atoxigenic strain. *Phytopathology* **87**: 940–945.

52. Keller, N.P., T.E. Cleveland & D. Bhatnagar. 1992. Variable electrophoretic karyotypes of members of *Aspergillus* Section *Flavi*. *Curr. Genet.* **21**: 371–375.

53. Egel, D.S., P.J. Cotty & K.S. Elias. 1994. Relationships among isolates of *Aspergillus* sect. *flavi* that vary in aflatoxin production. *Phytopathology* **84**: 906–912.

54. McAlpin, C.E., D.T. Wicklow & B.W Horn. 2002. DNA fingerprinting analysis of vegetative compatibility groups in *Aspergillus flavus* from a peanut field in Georgia. *Plant Dis.* **86**: 254–258.

55. Moody, S.F. & B.M. Tyler. 1990. Restriction enzyme analysis of mitochondrial DNA of the *Aspergillus flavus* group: *A. flavus*, *A. parasiticus*, and *A. nomius*. *Appl. Environ. Microbiol.* **56**: 2441–2452.

56. Moody, S.F. & B.M. Tyler. 1990. Use of nuclear DNA restriction fragment length polymorphisms to analyze the diversity of the *Aspergillus flavus* group: *A. flavus*, *A. parasiticus*, and *A. nomius*. *Appl. Environ. Microbiol.* **56**: 2453–2461.

57. Grubisha, L.C. & P.J. Cotty. 2009. Twenty-four microsatellite markers for the aflatoxin-producing fungus *Aspergillus flavus*. *Mol. Ecol. Resour.* **9**: 264–267.

58. Grubisha, L.C. & P.J. Cotty. 2010. Genetic isolation among sympatric vegetative compatibility groups of the aflatoxin-producing fungus *Aspergillus flavus*. *Mol. Ecol.* **19**: 269–280.

59. Ehrlich, K.C., B.G. Montalbano & P.J. Cotty. 2007. Analysis of single nucleotide polymorphisms in three genes shows evidence for genetic isolation of certain *Aspergillus flavus* vegetative compatibility groups. *FEMS Microbiol. Lett.* **268**: 231–236.

60. Geiser, D.M., J.I. Pitt & J.W. Taylor. 1998. Cryptic speciation and recombination in the aflatoxin-producing fungus *Aspergillus flavus*. *Proc. Natl. Acad. Sci. USA* **95**: 388–393.

61. Payne, G.A., W.C. Nierman, J.R. Wortman, *et al.* 2006. Whole genome comparison of *Aspergillus flavus* and *A. oryzae*. *Med. Mycol.* **44**: S9–S11.

62. Leslie, J.F. 1993. Fungal vegetative compatibility. *Annu. Rev. Phytopathol.* **31**: 127–150.

63. Papa, K.E. 1986. Heterokaryon incompatibility in *Aspergillus flavus*. *Mycologia* **78**: 98–101.

64. Bayman, P. & P.J. Cotty. 1993. Genetic diversity in *Aspergillus flavus*: association with aflatoxin production and morphology. *Can. J. Bot.* **71**: 23–34.

65. Horn, B.W., R.L. Greene, V.S. Sobolev, *et al.* 1996. Association of morphology and mycotoxins production with vegetative compatibility groups in *Aspergillus flavus*, *A. parasiticus*, and *A. tamarii*. *Mycologia* **88**: 574–587.

66. Novas, M.V. & D. Cabral. 2002. Association of mycotoxin and sclerotia production with compatibility groups in *Aspergillus flavus* from peanut in Argentina. *Plant Dis.* **86**: 215–219.

67. Pildain, M.B., G. Vaamonde & D. Cabral. 2004. Analysis of population structure of *Aspergillus flavus* from peanut based on vegetative compatibility, geographic origin, mycotoxin and sclerotia production. *Int. J. Food Microbiol.* **93**: 31–40.

68. Barros, G., A. Torres, M. Rodriguez & S. Chulze. 2006. Genetic diversity within *Aspergillus flavus* strains isolated from the peanut-cropped soils in Argentina. *Soil Biol. Biochem.* **38**: 145–152.

69. Barros, G.G., M.L. Chiotta, M.M. Reynoso, *et al.* 2007. Molecular characterization of *Aspergillus* section *Flavi* isolates collected from peanut fields in Argentina using AFLPs. *J. Appl. Microbiol.* **103**: 900–909.

70. Bayman, P. & P.J. Cotty. 1991. Vegetative compatibility and genetic diversity in the *Aspergillus flavus* population of a single field. *Can. J. Bot.* **69**: 1707–1711.

71. Horn, B.W. & R.L. Greene. 1995. Vegetative compatibility within populations of *Aspergillus flavus*, *A. parasiticus*, and *A. tamarii* from a peanut field. *Mycologia* **87**: 324–332.

72. Donner, M., J. Atehnkeng, R.A. Sikora, *et al.* 2009. Distribution of *Aspergillus* section Flavi in soils of maize fields in three agroecological zones of Nigeria. *Soil Biol. Biochem.* **41**: 37–44.

73. Horn, B.W. & J.W. Dorner. 1998. Soil populations of *Aspergillus* species from section *Flavi* along a transect through peanut-growing regions of the United States. *Mycologia* **90**: 767–776.

74. Jaime-Garcia, R. & P.J. Cotty. 2006. Spatial relationships of soil texture and crop rotation to *Aspergillus flavus* community structure in South Texas. *Phytopathology* **96**: 599–607.

75. Orum, T.V., D.M. Bigelow, P.J. Cotty & M.R. Nelson. 1999. Using predictions based on geostatistics to monitor trends in *Aspergillus flavus* strain composition. *Phytopathology* **89**: 761–769.

76. Fedorova, N.D., S. Harris, D. Chen, *et al.* 2009. Using aCGH to study intraspecific genetic variability in two pathogenic molds, *Aspergillus fumigatus* and *Aspergillus flavus*. *Med. Mycol.* **47**: S34–S31.

77. Horn, B.W, G.G. Moore & I. Carbone. 2009. Sexual reproduction in *Aspergillus flavus*. *Mycologia* **101**: 423–429.

78. Olarte, R.A., B.W. Horn, J.W. Dorner, *et al.* 2011. Effect of sexual recombination on population diversity in aflatoxin production by *Aspergillus flavus* and evidence for cryptic heterokaryosis. *Mol. Ecol.* **21**: 1453–1476.

79. Kwon-Chung, K.J. & J.A. Sugui. 2009. Sexual reproduction in *Aspergillus* species of medical or economical importance: why so fastidious? *Trends Microbiol.* **17**: 481–487.

80. Mellon, J.E., P.J. Cotty & M.K. Dowd. 2007. *Aspergillus flavus* hydrolases: their roles in pathogenesis and substrate utilization. *Appl. Microbiol. Biotechnol.* **77**: 497–504.

81. St. Leger, R.J., S.E. Screen & B. Shams-Pirzadeh. 2000. Lack of host specialization in *Aspergillus flavus*. *Appl. Environ. Microbiol.* **66**: 320–324.

82. Cotty, P.J., T.E. Cleveland, R.L. Brown & J.E. Mellon. 1990. Variation in polygalacturonase production among *Aspergillus flavus* isolates. *Appl. Environ. Microbiol.* **56:** 3885–3887.

83. Brown, R.L., T.E. Cleveland, P.J. Cotty & J.E. Mellon. 1992. Spread of *Aspergillus flavus* in cotton bolls, decay of intercarpellary membranes, and production of fungal pectinases. *Phytopathology* **82:** 462–467.

84. Cleveland, T.E. & P.J. Cotty. 1991. Invasiveness of *Aspergillus flavus* in wounded cotton bolls is associated with production of a specific fungal polygalacturonase. *Phytopathology* **81:** 155–158.

85. Shieh, M.T., R.L. Brown, M.P. Whitehead, *et al.* 1997. Molecular genetic evidence for the involvement of a specific polygalacturonase, P2c, in the invasion and spread of *Aspergillus flavus* in cotton bolls. *Appl. Environ. Microbiol.* **63:** 3548–3552.

86. Bilgrami, K.S. & A.K. Choudhary. 1993. Impact of habitats on toxigenic potential of *Aspergillus flavus*. *J. Stored Prod. Res.* **29:** 351–355.

87. Sweany, R.R., K.E. Damann & M.D. Kaller. 2011. Comparison of soil and corn kernel *Aspergillus flavus* populations: evidence for niche specialization. *Phytopathology* **101:** 952–959.

88. Mehl, H.L. & P.J. Cotty. 2011. Influence of the host contact sequence on the outcome of competition among *Aspergillus flavus* isolates during host tissue invasion. *Appl. Environ. Microbiol.* **77:** 1691–1697.

89. Schroeder, H.W. & R.A. Boller. 1973. Aflatoxin production of species and strains of the *Aspergillus flavus* group isolated from field crops. *Appl. Microbiol.* **25:** 885–889.

90. Vaamonde, G., A. Patriarca, V.F. Pinto, *et al.* 2003. Variability of aflatoxin and cyclopiazonic acid production by *Aspergillus* section *Flavi* from different substrates in Argentina. *Inter. J. Food Microbiol.* **88:** 79–84.

91. Jaime-Garcia, R. & P.J. Cotty. 2010. Crop rotation and soil temperature influence the community structure of *Aspergillus flavus* in soil. *Soil Biol. Biochem.* **42:** 1842–1847.

92. Fitt, B.D.L., Y.-J. Huang, F. van den Bosch & J.S. West. 2006. Coexistence of related pathogen species on arable crops in space and time. *Annu. Rev. Phytopathol.* **44:** 163–182.

93. Griffin, G.J., K.H. Garren & J.D. Taylor. 1981. Influence of crop rotation and minimum tillage on the population of *Aspergillus flavus* group in peanut field soil. *Plant Dis.* **65:** 898–900.

94. McGee, D.C., O.M. Olanya, G.M. Hoyos & L.H. Tiffany. 1996. Populations of *Aspergillus flavus* in the Iowa cornfield ecosystem in years not favorable for aflatoxin contamination of corn grain. *Plant Dis.* **80:** 742–746.

95. Cotty, P.J. & R. Jaime-Garcia. 2007. Influence of climate on aflatoxin producing fungi and aflatoxin contamination. *Int. J. Food Microbiol.* **119:** 109–115.

96. Cole, R.J., T.H. Sanders, R.A. Hill & P.D. Blankenship. 1985. Mean geocarposphere temperatures that induce preharvest aflatoxin contamination of peanuts under drought stress. *Mycopathologia* **91:** 41–46.

97. Payne, G.A., D.L. Thompson, E.B. Lillehoj, *et al.* 1988. Effect of temperature on the preharvest infection of maize kernels by *Aspergillus flavus*. *Phytopathology* **78:** 1376–1380.

98. Jaime, R., M. Foley, L. Antilla & P.J. Cotty. 2012. Long-term and area-wide influences of atoxigenic strain biocontrol technology for aflatoxin contamination. *Phytopathology* **102**(Suppl): S4.58.

Ann. N.Y. Acad. Sci. ISSN 0077-8923

ANNALS OF THE NEW YORK ACADEMY OF SCIENCES
Issue: *Advances Against Aspergillosis*

Current section and species complex concepts in *Aspergillus:* recommendations for routine daily practice

Ana Alastruey-Izquierdo, Emilia Mellado, and Manuel Cuenca-Estrella

Mycology Department, Spanish National Center for Microbiology, Instituto de Salud Carlos III, Madrid, Spain

Address for correspondence: Manuel Cuenca-Estrella, Servicio de Micología, Centro Nacional de Microbiología, Instituto de Salud Carlos III, Ctra Majadahonda-Pozuelo Km 2. 28220 Majadahonda (Madrid), Spain. mcuenca-estrella@isciii.es

The identification of fungi by molecular techniques has generated important changes in fungal taxonomy. The use of molecular tools in taxonomic studies has led to the description of some cryptic species that were placed into a complex of morphologically similar organisms and were subsequently misidentified. There are still limited data available on the prevalence of cryptic species of *Aspergillus* in the clinical setting, although some studies report 10–14% of the total number of *Aspergillus* species. In addition, the main concern about the emergence of *Aspergillus* cryptic species is that they can be more resistant to antifungal agents. The rise in the incidence of fungal infections and the changing landscape of epidemiology, together with the description of new pathogens and their different susceptibility profiles, make the identification by molecular methods and/or antifungal susceptibility testing the best options available for the correct management of these infections.

Keywords: *Aspergillus*; resistance; molecular identification

Introduction

The population at risk for invasive fungal infections is expected to increase in the coming years as new and evolving predisposing factors continue to rise. Despite advances in diagnosis and treatment, mortality and morbidity rates remain high. Although *Candida albicans* and *Aspergillus fumigatus* are the main etiologic agents of disease, the epidemiology of fungal infections is changing and the number of species able to infect humans is constantly growing.[1–4] This change in the epidemiology accounts for both yeast and molds. However, while the incidence of yeast has remained just about constant, in recent years an increase in the incidence of mold infections has been documented.[5,6] Furthermore, in some groups of patients molds are more frequently encountered than yeast.[6] The reasons for the changing epidemiology have been attributed to several reasons, such as the growing number of people at risk with impaired immunity, extensive use of antifungal agents (producing a shift toward more resistant organisms), and advances in diagno-sis and detection methods that have generated tools to achieve more accurate identification of fungal pathogens.

While yeasts are quite easily identified by their biochemical properties, mold identification in clinical laboratories relies mainly on the observation of morphological characteristics. For years the identification and the description of fungal species have been based on the observation of phenotypic characters. Morphological, and in some cases biochemical, properties were used to separate species. With the arrival of the genomic era the identification of fungi by molecular techniques generated (and still does) important changes in their taxonomy. Although there is no consensus for species delimitation in fungi, the phylogenetic species recognition (PSR) concept[7] has been demonstrated to be useful in several taxonomic studies.[2,3,8,9] This method is based on the sequencing of several targets to perform phylogenetic analysis for species recognition. In fact, based on PSR, new cryptic and sibling (from hereon "cryptic") species within the most frequent pathogens have been described.[2,3,10] Furthermore,

doi: 10.1111/j.1749-6632.2012.06822.x

other species have been reclassified under a different genus.[8,11]

The internal transcribed spacer (ITS) regions of ribosomal DNA (rDNA) have been the gold standard for identifying pathogenic fungi.[12] However, in some genera, such as *Fusarium, Scedosporium,* and *Aspergillus,* ITS regions identify only to the species "complex level" (that is, species that are morphologically or biochemically similar and otherwise indistinguishable by classical methods and/or ITS sequencing). Therefore, other targets have been selected as the preferred marker for classification purposes. Some of these new species are more resistant to current antifungal drugs, making the identification to species level of the etiologic agent of an infection decisive for the correct management of the patients. In this paper we will focus on the identification and susceptibility profile of the *Aspergillus* spp. isolates found in the clinical setting.

Aspergillus species in the clinical setting

Aspergillus spp. are the most frequent molds isolated from clinical samples. The genus *Aspergillus* contains about 250 species divided into eight subgenera (*Aspergillus, Fumigati, Circumdati, Candidi, Terrei, Nidulantes, Warcupi,* and *Ornati*),[13] which in turn are subdivided into several sections or species complexes. The most frequent cause of aspergillosis is *A. fumigatus,* followed by *A. flavus, A. terreus,* and *A. niger,* but many other species have been described in human infections. *A. terreus* has been particularly associated with lethal infections,[14,15] and *A. flavus* has been described as the most common species of *Aspergillus* isolated in some centers.[16]

The use of molecular tools in taxonomic studies has led to the description of some cryptic species that were misidentified and underestimated, as they were classified as complexes of morphologically similar species.[17] Until now, the consensus is that ITS sequencing allows classifying an isolate of *Aspergillus* within its species complex.[12] However, to reach a correct identification to a species level, the sequence of other targets, such as β-tubulin, calmodulin, or rodlet A genes,[17,18] has become necessary.

Taxonomic studies using molecular identification have produced important changes in the classification of *Aspergillus* species, cryptic species have been described and studies to reclassify strains have been conducted.[1,19–21] Nevertheless, the real distribution of these cryptic species in clinical samples remains

unclear. In a prospective study of fungal infections in transplant patients in the U. S.,[22] 218 *Aspergillus* isolates were identified. Using molecular identification, 147 (67.4%) were classified as *A. fumigatus* complex, 29 (13.2%) as *A. flavus* complex, 19 (8.7%) as *A. niger* complex, 11 (7.4%) as *A. terreus* complex, 6 (2.7%) as *A. ustus* complex, 5 (2.3%) as *A. versicolor* complex, and 1 as *A. nidulans* complex. Comparable results have been found in a prospective study performed in Spain (unpublished data) with samples from deep body sites. From a total of 278 *Aspergillus* isolates, 162 (50.3%) belonged to the *A. fumigatus* complex, 43 isolates (13.4%) to the *A. nigri* complex, 30 (9.3%) to the *A. flavus* complex, 27 (8.3%) to the *A. terreus* complex, 8 (2.9%) to the *A. nidulans* complex, 6 (2.2%) to the *A. usti* complex, and 1 to *A. versicolor* complex. Table 1 shows the distribution of species in both studies. Although the frequency and distribution of species are comparable between works, some differences can be noted. Thus, *Neosartoria udagawae* and *A. versicolor* were only identified in the U.S. study, while *A. nidulans* sensu stricto was only found in the Spanish study. In addition, the frequency of *A. tubingensis* and *A. terreus* was significantly lower in the U.S. study. These differences can be attributed to the study design, since the U.S. study included transplant patients only[22] and in the Spanish study samples were from patients with diverse underlying conditions; however, more studies are needed in order to know if these differences are stable and could be due to other factors such as differences in geographical distribution of *Aspergillus* species. In the U.S. study,[22] 10.6% (23 out of 218) isolates belonged to cryptic species of *Aspergillus,* and in the Spanish survey (unpublished data) up to 14.4% (40 out of 278) of the *Aspergillus* strains belonged to species that were not possible to identify by classical methods. Within the cryptic species, the most frequent ones were *A. tubingensis,* with 6 strains (2.8%) in the United States and 22 (7.9%) in Spain, *A. calidoustus,* with 6 isolates (2.8%) in the U. S. and 4 isolates (1.4%) in Spain, and *A. lentulus,* with 4 (1.8%) in the U. S. and 3 (1.1%) isolates in Spain. Other cryptic species such as *N. udagawae* and *A. alliaceus* were found only in one study (see Table 1). Many of these cryptic species have been previously described in human infections.[17,19,22–25] Moreover, some of them have been found to be in equal or greater frequency in clinical samples than their

Table 1. *Aspergillus* species distribution according to epidemiological surveys from Spain and the U. S.[22]

Species	Section	Transnet		Spanish study	
		N isolates	%	N isolates	%
A. fumigatus	Fumigati	139	63.8	156	56.1
A. lentulus	Fumigati	4	1.8	3	1.1
A. udagawae	Fumigati	3	1.4	0	0.0
N. peudofischeri	Fumigati	1	0.5	1	0.4
A. viridinutans	Fumigati	0	0.0	1	0.4
A. fumigatiafinis	Fumigati	0	0.0	1	0.4
A. flavus	Flavi	29	13.3	27	9.7
A. alliaceus	Flavi	0	0.0	3	1.1
A. terreus	Terrei	11	5.0	26	9.4
A. carneus	Terrei	0	0.0	1	0.4
A. tubingensis	Nigri	6	2.8	22	7.9
A. niger	Nigri	13	6.0	21	7.6
A. calidoustus	Usti	6	2.8	4	1.4
A. insuetus	Usti	0	0.0	1	0.4
A. keveii	Usti	0	0.0	1	0.4
A. sydowii	Versicolores	2	0.9	1	0.4
A. versicolor	Versicolores	3	1.4	0	0.0
E. quadrilineata	Nidulantes	1	0.5	0	0.0
A. nidulans	Nidulantes	0	0.0	8	2.9
A. westerdijkiae	Circumdati	0	0.0	1	0.4
Total		218	100	278	100

well-known relatives,[1,4,20,21] as it was found in the Spanish study with more strains of *A. tubingensis* than of *A. niger* (Table 1). Thus, the use of molecular identification seems to become an important tool in the clinical setting.

Antifungal susceptibility

One of the main concerns about the emergence of the cryptic species is that they can be more resistant to the antifungal drugs. The EUCAST Subcommittee on AFST, based on epidemiological studies,[26] has proposed breakpoints for voriconazole and posaconazole with regard to *A. fumigatus* and *A. terreus*, amphotericin B (AMB) with regard to *A. fumigatus* and *A. niger*, and itraconazole with regard to *A. fumigatus*, *A. flavus*, *A. terreus*, and *A. nidulans*.[27] Thus, strains with MIC (minimal inhibitory concentration) values > 2 mg/L for AMB, itraconazole, and voriconazole have been defined as resistant, while for posaconazole MIC values > 0.25 mg/L are considered resistant, according

to the EUCAST methodology. Taking into account these breakpoints, resistance has been analyzed in several studies. Table 2 shows geometric mean, MIC_{50} (MIC causing inhibition of 50% of isolates), MIC_{90} (MIC causing inhibition of 90% of isolates), and the range for the *Aspergillus* sections to which the most common clinical isolates belong. As showed in Table 2, sections Flavi and Terrei show higher MICs values to AMB. The reduced susceptibility of *A. terreus* to amphotericin B is well known.[28] Infections caused by these fungi have been associated with a lower response rate and poorer outcomes.[29] The reduced susceptibility of Flavi can be attributed to *A. alliaceus* (Table 2), which has been described as resistant to this drug.[30,31] Reduced susceptibility to AMB has also been found in cryptic species such as *A. lentulus*,[19,32] *A. fumigatiaffinis*,[19,22] *A. niveus*,[33] *A. carneus*,[34] or *A. calidoustus*.[1] It should be noted that the prevalence of these cryptic species is very low (<5% of isolates). Table 3 shows geometric mean, MIC_{50},

Table 2. Geometric mean, range, MIC_{50} and MIC_{90} from the species isolated from clinical samples grouped in *Aspergillus* sections. Data from the collection of the Spanish National Centre for Microbiology. MIC values in mg/L

Section (N)		Amphotericin B	Itraconazole	Voriconazole	Posaconazole
Fumigati (161)	GM	0.28	0.18	0.51	0.05
	Range	0.06–16	0.06–1	0.12–2	0.015–0.5
	MIC_{50}	0.25	0.25	0.5	0.06
	MIC_{90}	0.5	0.25	1	0.12
Flavi (30)	GM	2.15	0.19	0.63	0.09
	Range	0.5–32	0.03–1	0.12–4	0.015–0.25
	MIC_{50}	1	0.25	0.5	0.12
	MIC_{90}	32	1	1	0.12
Terrei (27)	GM	1.59	0.13	0.95	0.05
	Range	0.5–8	0.06–0.25	0.5–2	0.03–0.12
	MIC_{50}	1	0.12	1	0.06
	MIC_{90}	4	0.25	2	0.12
Nigri (43)	GM	0.12	0.43	0.74	0.10
	Range	0.06–1	0.06–16	0.25–2	0.015–0.25
	MIC_{50}	0.12	0.5	1	0.12
	MIC_{90}	0.25	1	1	0.12
Nidulantes (8)	GM	1.53	0.12	0.21	0.08
	Range	0.12–32	0.06–0.5	0.12–1	0.03–0.12
	MIC_{50}	1.5	0.12	0.25	0.12
	MIC_{90}	16	0.25	0.5	0.12
Usti (6)	GM	0.56	4	5.66	3.17
	Range	0.25–1	1–16	4–16	2–16
	MIC_{50}	0.5	8	4	2
	MIC_{90}	1	16	16	16
Versicolores (1)	MIC	0.50	1.00	1.00	0.50
Circumdatii (1)	MIC	32.00	0.50	0.50	0.12
All	GM	0.39	0.21	0.60	0.07
	Range	0.06–32	0.015–16	0.12–16	0.015–16
	MIC_{50}	0.25	0.25	0.5	0.06
	MIC_{90}	2	0.5	1	0.12

N, number of isolates tested; GM, geometric mean of MICs; MIC_{50}, MIC causing inhibition of 50% of isolates; MIC_{90}, MIC causing inhibition of 90% of isolates.

MIC_{90}, and range for the most frequent cryptic species.

Reduced susceptibility to azole drugs have been extensively investigated in *A. fumigatus*.[35–38] Several reports describe an increasing incidence of strains with acquired resistance associated with an elevated mortality, but such emergence of resistance strains has only been found in some European countries.[39,40] On the other hand, some cryptic species within *A. fumigatus* species complex, such as *A. lentulus, A. viridinutans, N. pseudofischeri, A. fumigatiafinnis,* have been described to be resistant to all azoles.[19,22,39,40] This pattern of resistance has been also described in species from other complexes, such as *A. tubingensis* (*A. niger* complex)[19,21] and *A. calidoustus* (*A. ustus* complex) that have an azole multiresistant susceptibility profile.[1] *In vitro* susceptibility data for these species are shown in Table 3 (data from the Spanish National Centre for Microbiology).

Table 3. Geometric mean, range, MIC$_{50}$ and MIC$_{90}$ from most common cryptic species of *Aspergillus* with decreased susceptibility to antifungal agents. Data from the Spanish National Centre for Microbiology. MIC values are expressed as mg/L

Species (N)		Amphotericin B	Itraconazole	Voriconazole	Posaconazole
A. lentulus (17)	GM	6.40	3.57	3.22	0.20
	Range	1–16	0.25–16	2–8	0.06–2
	MIC$_{50}$	8	8	4	0.12
	MIC$_{90}$	16	16	8	2
A. alliaceus (25)	GM	21.71	0.21	0.54	0.15
	Range	0.5–32	0.125–16	0.25–2	0.06–16
	MIC$_{50}$	32	0.125	0.5	0.12
	MIC$_{90}$	32	2	2	0.5
A. tubingensis (22)	GM	0.09	0.51	0.78	0.10
	Range	0.06–1	0.06–16	0.25–2	0.015–0.25
	MIC$_{50}$	0.12	0.5	1	0.12
	MIC$_{90}$	0.12	1	2	0.12
A. calidoustus (12)	GM	0.80	8.31	5.88	7.41
	Range	0.5–2	1–16	4–8	2–16
	MIC$_{50}$	1	16	8	8
	MIC$_{90}$	2	16	8	16

N, number of isolates tested; GM, geometric mean of MICs; MIC$_{50}$, MIC causing inhibition of 50% of isolates; MIC$_{90}$, MIC causing inhibition of 90% of isolates.

Conclusions

Fungal infections are becoming more frequent in the clinical setting. Despite technical advances and the availability of more antifungal compounds, the survival rate among these patients remains low. Proper identification of the isolate and early initiation of an effective antifungal therapy are essential for patient recovery. Recent advances in molecular tools have allowed for the description of new cryptic species among different groups of fungi. This has been particularly significant in *Aspergillus* since new species are almost impossible to differentiate by classical morphological tools and some are more resistant to antifungal drugs. ITS sequencing allows classifying an isolate within its species complex but to reach a proper identification to species level it is necessary to sequence other targets such as beta tubulin. The frequency of these cryptic species is still not clear, some data report between 10 and 14% of the total *Aspergillus* strains. But more studies are warranted in order to elucidate the real prevalence of these species in the clinical setting.

The rise in the incidence of fungal infections, the changing landscape of epidemiology, together with the description of new pathogens and their different susceptibility profile, make the identification by molecular methods and/or antifungal susceptibility testing the best options available for the correct management of these infections.

The classification of isolates by molecular methods could be useful for the clinical management of patients, since over 10% of *Aspergillus* strains could belong to cryptic species, some of which are more resistant *in vitro* to antifungal agents. More studies should be done to understand the local epidemiology in each area.

Some recommendations can be made for routine daily practice. When dealing with strains on cultures from deep sites we recommend performing identification by β-tubulin sequencing. Where such sequencing is not available, other strategies for *A. fumigati* complex, such as matrix-assisted laser deroption/ionication/time of flight mass spectrometer[41] or luminex assay,[42] appear to be useful for identifying cryptic species. When other strategies are not available, we advise performing antifungal susceptibility testing of the strains to assess their susceptibility profile *in vitro* and then sending the strains to the reference centre for correct classification and epidemiological analysis. It has to be taken into account that although proper identification can speed up directed antifungal therapy, these techniques cannot detect strains with secondary resistance. Therefore, susceptibility testing is an essential tool and should be performed routinely for better management of antifungal therapy.

Acknowledgments

A.A.-I. has a research contract from REIPI (Red Española de Investigación en Patología Infecciosa, Project MPY 1022/07_1), and from Instituto de Salud Carlos III – cofinanced by European Development Regional Fund "A way to achieve Europe" ERDF, Spanish Network for the Research in Infectious Diseases (REIPI RD06/0008).

Conflicts of interest

M.C.E. has received grant support from Astellas Pharma, BioMerieux, Gilead Sciences, Merck Sharp and Dohme, Pfizer, Schering Plough, Soria Melguizo SA, Ferrer International the European Union, the ALBAN program, the Spanish Agency for International Cooperation, the Spanish Ministry of Culture and Education, The Spanish Health Research Fund, The Instituto de Salud Carlos III, The Ramon Areces Foundation, and The Mutua Madrilena Foundation. He has been an advisor/consultant to the Panamerican Health Organization, Gilead Sciences, Merck Sharp and Dohme, Pfizer, and Schering Plough. He has been paid for talks on behalf of Gilead Sciences, Merck Sharp and Dohme, Pfizer, and Schering Plough.

References

1. Alastruey-Izquierdo, A., I. Cuesta, J. Houbraken, *et al.* 2010. In vitro activity of nine antifungal agents against clinical isolates of Aspergillus calidoustus. *Med. Mycol.* **48:** 97–102.
2. Balajee, S.A., J.L. Gribskov, E. Hanley, *et al.* 2005. *Aspergillus lentulus* sp. nov., a new sibling species of *A. fumigatus*. *Eukaryot. Cell* **4:** 625–632.
3. Gilgado, F., J. Cano, J. Gene & J. Guarro. 2005. Molecular phylogeny of the *Pseudallescheria boydii* species complex: proposal of two new species. *J. Clin. Microbiol.* **43:** 4930–4942.
4. Varga, J., J. Houbraken, H.A. Van Der Lee, *et al.* 2008. *Aspergillus calidoustus* sp. nov., causative agent of human infections previously assigned to *Aspergillus ustus*. *Eukaryot. Cell* **7:** 630–638.
5. Kontoyiannis, D.P., K.A. Marr, B.J. Park, *et al.* 2010. Prospective surveillance for invasive fungal infections in hematopoietic stem cell transplant recipients, 2001–2006: overview

6. Lass-Florl, C. 2009. The changing face of epidemiology of invasive fungal disease in Europe. *Mycoses* **52:** 197–205.
7. Taylor, J.W., D.J. Jacobson, S. Kroken, *et al.* 2000. Phylogenetic species recognition and species concepts in fungi. *Fungal Genet. Biol.* **31:** 21–32.
8. Alastruey-Izquierdo, A., K. Hoffmann, G.S. De Hoog, *et al.* 2010. Species recognition and clinical relevance of the zygomycetous genus *Lichtheimia* (syn. *Absidia* pro parte, *Mycocladus*). *J. Clin. Microbiol.* **48:** 2154–2170.
9. O'Donnell, K., D.A. Sutton, M.G. Rinaldi, *et al.* 2009. Novel multilocus sequence typing scheme reveals high genetic diversity of human pathogenic members of the *Fusarium incarnatum-F. equiseti* and *F. chlamydosporum* species complexes within the United States. *J. Clin. Microbiol.* **47:** 3851–3861.
10. Balajee, S.A., J.W. Baddley, S.W. Peterson, *et al.* 2009. *Aspergillus alabamensis*, a new clinically relevant species in the section. *Terrei Eukaryot. Cell* **8:** 713–722.
11. Alvarez, E., J. Cano, A.M. Stchigel, *et al.* 2011. Two new species of *Mucor* from clinical samples. *Med. Mycol.* **49:** 62–72.
12. Balajee, S.A., A.M. Borman, M.E. Brandt, *et al.* 2009. Sequence-based identification of *Aspergillus*, *Fusarium*, and mucorales species in the clinical mycology laboratory: where are we and where should we go from here? *J. Clin. Microbiol.* **47:** 877–884.
13. Samson, R. & J. Varga. 2012. Molecular Systematics of *Aspergillus* and its Teleomorphs. In *Aspergillus: Molecular Biology and Genomics*. M. Machida & K. Gomi, Eds.: 19–40. Caister Academic Press. Wymondham, UK.
14. Hachem, R.Y., D.P. Kontoyiannis, M.R. Boktour, *et al.* 2004. *Aspergillus terreus*: an emerging amphotericin B-resistant opportunistic mold in patients with hematologic malignancies. *Cancer* **101:** 1594–1600.
15. Lass-Florl, C., G. Kofler, G. Kropshofer, *et al.* 1998. In-vitro testing of susceptibility to amphotericin B is a reliable predictor of clinical outcome in invasive aspergillosis. *J. Antimicrob. Chemother.* **42:** 497–502.
16. Hedayati, M.T., A.C. Pasqualotto, P.A. Warn, *et al.* 2007. *Aspergillus flavus*: human pathogen, allergen and mycotoxin producer. *Microbiology* **153:** 1677–1692.
17. Balajee, S.A., J. Gribskov, M. Brandt, *et al.* 2005. Mistaken identity: *Neosartorya pseudofischeri* and its anamorph masquerading as *Aspergillus fumigatus*. *J. Clin. Microbiol.* **43:** 5996–5999.
18. Hong, S.B., S.J. Go, H.D. Shin, *et al.* 2005. Polyphasic taxonomy of *Aspergillus fumigatus* and related species. *Mycologia* **97:** 1316–1329.
19. Alcazar-Fuoli, L., E. Mellado, A. Alastruey-Izquierdo, *et al.* 2008. *Aspergillus* section Fumigati: antifungal susceptibility patterns and sequence-based identification. *Antimicrob. Agents Chemother.* **52:** 1244–1251.
20. Hendrickx, M., H. Beguin & M. Detandt. 2012. Genetic re-identification and antifungal susceptibility testing of *Aspergillus* section Nigri strains of the BCCM/IHEM collection. *Mycoses* **55:** 148–155.

21. Howard, S.J., E. Harrison, P. Bowyer, *et al.* 2011. Cryptic species and azole resistance in the *Aspergillus niger* complex. *Antimicrob. Agents Chemother.* **55:** 4802–4809.

22. Balajee, S.A., R. Kano, J.W. Baddley, *et al.* 2009. Molecular identification of *Aspergillus* species collected for the Transplant-Associated Infection Surveillance Network. *J. Clin. Microbiol.* **47:** 3138–3141.

23. Gerber, J., J. Chomicki, J.W. Brandsberg, *et al.* 1973. Pulmonary aspergillosis caused by *Aspergillus fischeri* var. *spinosus*: report of a case and value of serologic studies. *Am. J. Clin. Pathol.* **60:** 861–866.

24. Guarro, J., E.G. Kallas, P. Godoy, *et al.* 2002. Cerebral aspergillosis caused by *Neosartorya hiratsukae*. *Brazil Emerg. Infect. Dis.* **8:** 989–991.

25. Jarv, H., J. Lehtmaa, R.C. Summerbell, *et al.* 2004. Isolation of *Neosartorya pseudofischeri* from blood: first hint of pulmonary Aspergillosis. *J. Clin. Microbiol.* **42:** 925–928.

26. Rodriguez-Tudela, J.L., L. Alcazar-Fuoli, E. Mellado, *et al.* 2008. Epidemiological cutoffs and cross-resistance to azole drugs in *Aspergillus fumigatus*. *Antimicrob. Agents Chemother.* **52:** 2468–2472.

27. AFST EUCAST. 2012. Clinical breakpoints – fungi (v 4.1). http://www.eucast.org/fileadmin/src/media/PDFs/EUCAST_files/AFST/Antifungal_breakpoints_v_4.1.pdf.

28. Lass-Florl, C., A. Alastruey-Izquierdo, M. Cuenca-Estrella, *et al.* 2009. In vitro activities of various antifungal drugs against *Aspergillus terreus*: global assessment using the methodology of the European committee on antimicrobial susceptibility testing. *Antimicrob. Agents Chemother.* **53:** 794–795.

29. Lass-Florl, C., K. Griff, A. Mayr, *et al.* 2005. Epidemiology and outcome of infections due to *Aspergillus terreus*: 10-year single centre experience. *Br. J. Haematol.* **131:** 201–207.

30. Balajee, S.A., M.D. Lindsley, N. Iqbal, *et al.* 2007. Nonsporulating clinical isolate identified as *Petromyces alliaceus* (anamorph *Aspergillus alliaceus*) by morphological and sequence-based methods. *J. Clin. Microbiol.* **45:** 2701–2703.

31. Ozhak-Baysan, B., A. Alastruey-Izquierdo, R. Saba, *et al.* 2010. *Aspergillus alliaceus* and *Aspergillus flavus* co-infection in an acute myeloid leukemia patient. *Med. Mycol.* **48:** 995–999.

32. Baddley, J.W., K.A. Marr, D.R. Andes, *et al.* 2009. Patterns of susceptibility of *Aspergillus* isolates recovered from patients enrolled in the Transplant-Associated Infection Surveillance Network. *J. Clin. Microbiol.* **47:** 3271–3275.

33. Auberger, J., C. Lass-Florl, J. Clausen, *et al.* 2008. First case of breakthrough pulmonary *Aspergillus niveus* infection in a patient after allogeneic hematopoietic stem cell transplantation. *Diagn. Microbiol. Infect. Dis.* **62:** 336–339.

34. Escribano, P., T. Pelaez, S. Recio, *et al.* 2012. Characterization of clinical strains of *Aspergillus terreus* complex: molecular identification and antifungal susceptibility to azoles and amphotericin. *B Clin. Microbiol. Infect.* **18:** E24–E26.

35. Mellado, E., T.M. Diaz-Guerra, M. Cuenca-Estrella & J.L. Rodriguez-Tudela. 2001. Identification of two different 14-alpha sterol demethylase-related genes (cyp51A and cyp51B) in *Aspergillus fumigatus* and other *Aspergillus* species. *J. Clin. Microbiol.* **39:** 2431–2438.

36. Mellado, E., G. Garcia-Effron, L. Alcazar-Fuoli, *et al.* 2004. Substitutions at methionine 220 in the 14alpha-sterol demethylase (Cyp51A) of *Aspergillus fumigatus* are responsible for resistance in vitro to azole antifungal drugs. *Antimicrob. Agents Chemother.* **48:** 2747–2750.

37. Mellado, E., G. Garcia-Effron, L. Alcazar-Fuoli, *et al.* 2007. A new *Aspergillus fumigatus* resistance mechanism conferring in vitro cross-resistance to azole antifungals involves a combination of cyp51A alterations. *Antimicrob. Agents Chemother.* **51:** 1897–1904.

38. Mellado, E., L. Alcazar-Fuoli, M. Cuenca-Estrella & J.L. Rodriguez-Tudela. 2011. Role of *Aspergillus lentulus* 14-alpha sterol demethylase (Cyp51A) in azole drug susceptibility. *Antimicrob. Agents Chemother.* **55:** 5459–5468.

39. Howard, S.J., D. Cerar, M.J. Anderson, *et al.* 2009. Frequency and evolution of Azole resistance in *Aspergillus fumigatus* associated with treatment failure. *Emerg. Infect. Dis.* **15:** 1068–1076.

40. Snelders, E., H.A. Van Der Lee, J. Kuijpers, *et al.* 2008. Emergence of azole resistance in *Aspergillus fumigatus* and spread of a single resistance mechanism. *PLoS Med.* **5:** e219.

41. Alanio, A., J.L. Beretti, B. Dauphin, *et al.* 2011. Matrix-assisted laser desorption ionization time-of-flight mass spectrometry for fast and accurate identification of clinically relevant *Aspergillus* species. *Clin. Microbiol. Infect.* **17:** 750–755.

42. Etienne, K.A., L. Gade, S.R. Lockhart, *et al.* 2009. Screening of a large global *Aspergillus fumigatus* species complex collection by using a species-specific microsphere-based Luminex assay. *J. Clin. Microbiol.* **47:** 4171–4172.

Ann. N.Y. Acad. Sci. ISSN 0077-8923

ANNALS OF THE NEW YORK ACADEMY OF SCIENCES

Issue: *Advances Against Aspergillosis*

The diverse applications of RNA-seq for functional genomic studies in *Aspergillus fumigatus*

Antonis Rokas,[1] John G. Gibbons,[1] Xiaofan Zhou,[1] Anne Beauvais,[2] and Jean-Paul Latgé[2]

[1]Department of Biological Sciences, Vanderbilt University, Nashville, Tennessee. [2]Unité des Aspergillus, Institut Pasteur, Paris, France

Address for correspondence: Antonis Rokas, Department of Biological Sciences, Vanderbilt University, VU Station B #35-1634, Nashville, TN 37235. antonis.rokas@vanderbilt.edu

The deep sequencing of an mRNA population, RNA-seq, is a very successful application of next-generation sequencing technologies (NGSTs). RNA-seq takes advantage of two key NGST features: (1) samples can be mixtures of different DNA pieces, and (2) sequencing provides both qualitative and quantitative information about each DNA piece analyzed. We recently used RNA-seq to study the transcriptome of *Aspergillus fumigatus*, a deadly human fungal pathogen. Analysis of the RNA-seq data indicates that there are likely tens of unannotated and hundreds of novel genes in the *A. fumigatus* transcriptome, mostly encoding for small proteins. Inspection of transcriptome-wide variation between two isolates reveals thousands of single nucleotide polymorphisms. Finally, comparison of the transcriptome profiles of one isolate in two different growth conditions identified thousands of differentially expressed genes. These results demonstrate the utility and potential of RNA-seq for functional genomics studies in *A. fumigatus* and other fungal human pathogens.

Keywords: novel genes; annotation; population structure; differential expression; transcriptome profiling

Introduction

Recent technological advances in genome science have enabled researchers to routinely generate unprecedented amounts of sequence data from almost any species, opening the floodgates for the study of genome content and function in non-model organisms.[1,2] The main catalyst for these changes has been the development of several different so-called next-generation sequencing technologies (NGSTs).[3] Astonishingly, the amount of sequence data that a single NGST machine can currently produce in a few days is larger than the total amount of sequence data ever collected and deposited in sequence databases by individual users through traditional methods.[4] Importantly, NGST technologies yield not only *qualitative* information about the sequence of every DNA fragment analyzed, but also *quantitative* information about the relative abundance of each DNA fragment in the library sequenced.[1]

The abundance of NGST-produced data, their qualitative and quantitative nature, and their applicability to any organism for which fresh DNA or RNA is available, has enabled researchers to tailor NGSTs for a variety of different questions beyond the sequencing of genomes.[1,5] One of the most powerful such applications is RNA-seq, the employment of NGSTs for transcriptome profiling.[6,7] A typical RNA-seq experiment begins with the isolation of mRNA, its conversion into cDNA, followed by fragmentation and addition of adaptors to each DNA fragment's ends (Fig. 1). Sequencing the library of fragments in a high-throughput fashion returns as many as billions of sequence reads that can vary in length and in their characteristics (e.g., single-end or paired-end), depending on the NGST technology being employed and the experiment being conducted.[3] Once sequence reads have been obtained, they can be used for a wide variety of functional analyses, including, but not limited to, the study of alternative splicing, gene expression,

doi: 10.1111/j.1749-6632.2012.06755.x

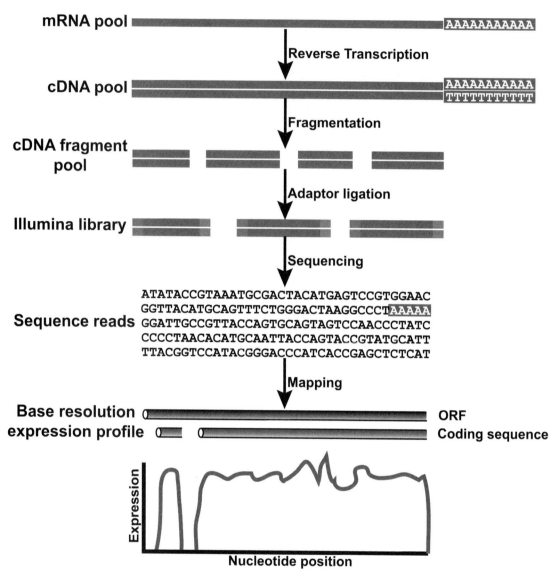

Figure 1. The workflow of a typical RNA-seq experiment. In brief, following isolation, the mRNA pool is converted to a cDNA pool and is then fragmented. Next, NGST adaptors are added to each cDNA fragment, the resulting library of fragments is sequenced, and the sequence of each fragment is read using NGST. Once sequence reads have been obtained, they can be used for a variety of analyses. For example, if the mRNA pool is from an organism whose genome and annotation is known, the sequence reads can be aligned or mapped to the reference genome or transcriptome and the sequence of the entire transcript as well as its relative expression can be calculated. Thus, RNA-seq technology is simultaneously *qualitative* (i.e., it can determine the sequences of different sequence fragments in a pool) and *quantitative* (i.e., it can determine the relative abundance of different sequence fragments in a pool).

allele-specific expression, identification of transcription start sites, identification of gene fusion,[6,8,9] as well as for a variety of evolutionary analyses.[2,10] For example, in what is perhaps its most frequent application, when RNA-seq is performed on an organism whose genome and annotation is already char-

acterized, one can directly map the sequence reads to the reference genome or transcriptome, thus simultaneously calculating its abundance as well as its sequence (Fig. 1).

The filamentous fungal genus *Aspergillus* contains approximately 250 species and spans over

200 million years of evolutionary history.[11] Several species in the genus can cause a range of frequently deadly diseases, which are collectively known as aspergillosis.[12,13] Aspergillosis usually affects individuals that have compromised immune defenses and is established following inhalation of *Aspergillus* spores. The great majority of *Aspergillus*-induced infections is caused by *Aspergillus fumigatus*,[12,14,15] a very abundant and widely distributed species. *A. fumigatus* is one of the most common species found in decaying vegetation, and a prolific spore producer.[16–19] When *A. fumigatus* establishes an infection in the human lung, it usually forms a dense colony of filaments embedded in a polymeric extracellular matrix.[20,21] To identify candidate genes involved in this colony or biofilm-like growth (COG), we previously used RNA-seq to compare the transcriptomes of COG and liquid planktonic growth (PLG) conditions.[22] Here, we use this data to highlight the multitude of utilities of RNA-seq technology for functional genomics studies of *A. fumigatus*, by focusing specifically on three applications: characterizing the structure of the *A. fumigatus* transcriptome; measuring transcriptome-wide levels of variation between isolates; and, finally, comparing the transcriptome-wide expression profile of *A. fumigatus* during different nonnutritional growth conditions.

RNA-seq for annotation: characterizing transcriptome structure

Our RNA-seq data provided very good coverage of the *A. fumigatus* Af293 reference transcriptome. Specifically, 27,236,154 sequence reads, each 42 bp long and generated by RNA-seq from growth of the ATCC46645 isolate in the COG and PLG conditions, were mapped against the Af293 reference transcriptome. Sequence reads mapped to 90.5%

(8,952/9,887) of reference transcripts, recovering 77% of the sequence of the reference transcriptome (11.1/14.4 Mb). Approximately 73% of the sequence of the average transcript was recovered but for over 60% of the transcripts, more than 90% of their sequence was recovered.

To identify the extent of unannotated and novel genes in the *A. fumigatus* genome, we further mapped the RNA-seq generated sequence reads to the *A. fumigatus* Af293 reference genome, using reference gene models[23] as guides. In brief, gene models are hypotheses about the structure of transcripts produced by the set of genes in a genome. Although the majority of gene models constructed through the annotation of the reference Af293 isolate is of high quality and supported by a variety of evidence (e.g., expressed sequence tags and high similarity scores to other genes), annotation is a very challenging process and examples of misannotated genes or genes omitted from an annotation are present in any eukaryotic genome.[24] After the mapping step, we assembled the mapped reads into gene models using two different state-of-the-art programs, Cufflinks[25] and Scripture.[26] Interestingly, whereas the Cufflinks program is designed to maximize precision, the Scripture program is designed to maximize sensitivity, resulting in significantly different annotations from the same set of data, especially for lower-expressed genes.[9] We then compared the annotation produced by the two programs to the reference annotation and on the basis of the results classified the gene models constructed by the two programs as annotated, unannotated, or novel. Gene models whose sequence overlapped with that from any reference gene and whose location was on the same strand as their reference counterpart were considered to be "annotated." For remaining gene models, we used their protein sequence

Table 1. Summary of unannotated and novel genes reconstructed using RNA-seq data

| Software | Annotated genes | Unannotated genes | | Novel genes |
		Protein-coding	RNA-coding	
Cufflinks	7,764 (8,663)	390 (411)	24 (25)	1,673 (1,703)
Scripture	4,285 (6,852)	156 (214)	23 (54)	462 (653)

NOTE: Number in bracket indicates the total number of transcripts in each category. Some loci are included in more than one category.

products in BLAST similarity searches against the NCBI *nr* database for the presence of homologs in the genome of any organism. Gene models with significant hit(s) in the *nr* database were considered "unannotated," whereas gene models without any hit(s) in the same database were considered "novel."

Examination of the annotation produced from the Cufflinks and Scripture programs using the RNA-seq data identified hundreds of unannotated and thousands of novel gene models (Table 1). In line with previous analyses,[9] the numbers of unannotated and novel gene models differed greatly between the two programs. Interestingly, the great majority of novel genes predicted by the two programs encoded for small proteins (Fig. 2), with 91% (1,519/1,673) and 83% (382/460) of the protein products of novel gene models constructed by Cufflinks and Scripture, respectively, being equal or shorter than 120 amino acids (aa). In contrast, the percentages for unannotated genes were 45% (Cufflinks) and 43% (Scripture). Similarly, the median lengths of novel proteins were 68 aa (Cufflinks) and 72 aa (Scripture), whereas the median lengths of

unannotated proteins were 132 aa (Cufflinks) and 136 aa (Scripture). Although these analyses are preliminary, the RNA-seq data suggest that, conservatively, there are likely tens of previously unannotated and hundreds of novel gene models in *A. fumigatus*, the great majority of which encode for small proteins. Determining the function of these small proteins as well as their role in pathogenicity, if any, is a very interesting research challenge and opportunity for future functional genomics research in *A. fumigatus*.

RNA-seq for population genetics: characterizing transcriptome variation between isolates

Although many fungal species, including several human pathogens, show population structure,[28–30] it was thought that *A. fumigatus* lacks population structure.[31–33] However, two recent multilocus studies, one using isolates from around the world[32] and the other using isolates from the Netherlands,[34] identified genetically distinct lineages within *A. fumigatus*, suggesting that the absence of population structure in older studies could be due to

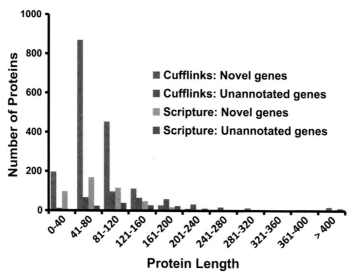

Figure 2. The majority of putative novel genes identified by RNA-seq in *A. fumigatus* encodes for small proteins. ATCC46645 sequence reads were mapped against the *A. fumigatus* Af293 reference genome with the TopHat software,[27] using the reference gene models[23] as guides and not allowing for introns >1,000 bp. Mapped reads were assembled into transcripts using the Cufflinks[25] and Scripture[26] programs, and their gene predictions were compared to the reference gene models. ORFs overlapping with exons of any reference gene model were classified as "annotated" genes. For the remaining ORFs, their protein products were searched against the NCBI *nr* database; these loci were classified either as "unannotated," if the encoded protein had at least one homolog in the NCBI *nr* database, or as "novel," if the encoded protein had no homologs in the NCBI *nr* database. The *x*-axis corresponds to groups of protein lengths encoded by putative novel or unannotated genes identified by the Cufflinks and Scripture programs (indicated by bars of different colors). The *y*-axis corresponds to the number of genes belonging to each group.

the use of fewer and less informative markers. To evaluate the potential of RNA-seq to provide novel markers for population genetic and, more generally, evolutionary analysis, we compared the transcriptomes of ATCC46645 (reconstructed through mapping to the Af293 reference transcriptome) and Af293 isolates. Examination of sequence alignments between the two isolates from 8,952 genes identified 12,872 single nucleotide polymorphisms (SNPs) in 4,923 genes, representing nearly 50% (4,923/9,887) of *A. fumigatus* reference genes. The average number of SNPs in variable genes was 2.6, with nearly two-thirds of the genes containing 1 or 2 SNPs, 12 containing more than 10 SNPs, and two containing 40 SNPs. Per kilobase of transcriptome sequence, the average SNP density was 1.2 (12,872 SNPs/11,109,536 recovered nucleotides; Fig. 3A), with 12 genes showing SNP densities greater than 10 (Table 2 and Fig. 3B). Finally, 51% of SNPs were nonsynonymous substitutions, with the remaining 49% being synonymous ones.

The identification of thousands of SNPs between the two isolates, suggests that in addition to whole-genome sequencing, RNA-seq is a powerful tool for the study of genetic differentiation in *A. fumigatus*. Importantly, because the *A. fumigatus* transcriptome is approximately 50% of the genome and because transcripts are grossly unevenly abundant

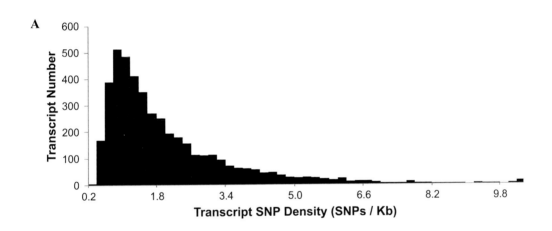

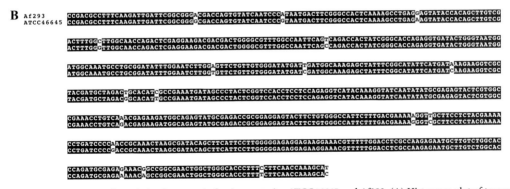

Figure 3. Transcriptome-wide variation between *A. fumigatus* strains ATCC46645 and Af293. (A) Histogram plot of transcript SNP density. The *x*-axis corresponds to groups of different transcript SNP densities (number of SNPs/Kb). The *y*-axis corresponds to the number of transcripts in each SNP density group. (B) Partial nucleotide sequence alignment between Af293 and ATCC46645 strains for a putative protein kinase (Afu3g03740), one of the genes with the highest SNP density in our comparison (see also Table 2). High-quality SNPs were identified by filtering for variable sites with coverage values ≥ 5 and an average quality scores ≥ 20 using the cns2snp script in the Maq software.[35] Pairwise nucleotide diversity was calculated as $\pi = n/N$, where n is the number of differences between sequences, and N is the total number of sites examined. SNP density per Kb was calculated as $\pi * 1,000$.

Table 2. Transcripts with SNP densities greater than 10 SNPs per kilobase

Gene name	Function	SNP density (total SNPs)
Afu2g01890	CAT5 protein	10.4 (8)
Afu2g17900	Conserved hypothetical protein	14.3 (7)
Afu3g01500	Hypothetical protein	10.3 (7)
Afu3g03740	Putative protein kinase	27.0 (18)
Afu3g07310	Conserved hypothetical protein	14.5 (12)
Afu4g00580	Hypothetical protein	10.1 (10)
Afu5g00890	Hypothetical protein	10.2 (2)
Afu5g01650	Putative bZIP transcription factor (JlbA)	10.4 (7)
Afu6g03420	Clock-controlled gene-9 protein	11.4 (7)
Afu7g08410	Putative transposase	19.3 (18)
Afu7g08470	Peroxisomal copper amine oxidase	11.0 (18)
Afu8g06430	Hypothetical protein	12.2 (3)

(varying over several orders of magnitude[22]), even shallow NGST sequencing should provide in depth sampling of a few hundred loci, and of hundreds if not thousands of SNPs, simply by sequencing transcripts in proportion to their representation in the library.[10] Thus, RNA-seq is a powerful alternative to the standard multilocus sequence typing currently used for the study of isolate identification and population structure in *A. fumigatus*, and in filamentous fungi in general.[36]

RNA-seq for functional genomics: characterizing global transcriptome changes between different growth conditions

The most common application of RNA-seq is for the identification of genes that show differential regulation under certain conditions. Examination of gene expression using our RNA-seq data obtained from *A. fumigatus* growth in the COG and PLG conditions revealed that 92% of reference transcripts (9,099/9,887) were expressed in both conditions and 4.3% (426/9,887) were uniquely expressed in either condition. By considering differentially expressed genes as only those that exhibited a twofold biological difference in relative gene expression between

conditions and a statistically significant *P* value below 5.5e–06, we identified 2,861 genes that were either significantly upregulated in COG relative to PLG or uniquely expressed in COG, and 1,339 that were either significantly downregulated in the same comparison or uniquely expressed in PLG (Fig. 4A). The remaining genes either showed uniform expression in the two conditions (5,370) or were not expressed in either condition (362; Fig. 4B). Remarkably, the range of expression values in both samples ranged seven orders of magnitude.

Upregulated and downregulated genes were nonrandomly distributed across the genome and showed strong association with specific functional categories (Fig. 5).[22] Some of the strongest associations were for cell wall genes (of 409 genes, 169 were significantly upregulated and only 41 were significantly downregulated), for pump and transporter genes (of 319 genes, 146 were upregulated and 16 were downregulated), and for allergens (81 genes, 41 were upregulated and 13 were downregulated). Thus, the application of RNA-seq to study a single difference in nonnutritional environmental conditions identified thousands of differentially expressed genes. Considering that the breadth and sensitivity of other technologies for measuring macromolecule abundance differences, such as microarrays and 2D gel electrophoresis, are much narrower,[37] RNA-seq appears to be the most powerful tool for genome-wide functional comparisons of fungal growth to date.

Finally, it is important to emphasize that the gene expression values measured by RNA-seq do not show significant variation when replicated. Biological and technical replicates are not yet standard in the RNA-seq literature, for the simple reason that the RNA-seq technique is much more accurate than microarrays,[38] although the inclusion of technical and biological replicates offers additional power.[39] We performed both biological and technical replicates for a subset of our RNA-seq experiments to verify that our RNA-seq experiments worked as expected.[22] We observed a very high degree of replicability of our results at both the biological and the technical level in the subset tested. For example, the correlation values between three biological PLG replicates performed on three different *A. fumigatus* strains as well as on one technical BFG replicate are extremely high ($r > 0.91$),[22] on par with similar studies in the literature.[40] These data suggest that

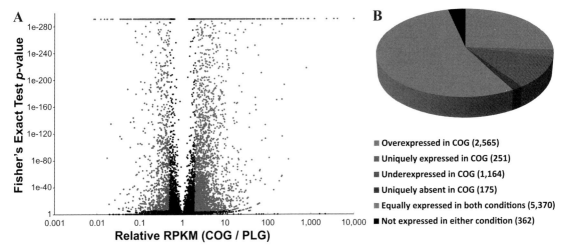

Figure 4. RNA-seq identifies thousands of differentially expressed genes from *A. fumigatus* ATCC46645 grown in the colony (COG) and plankton (PLG) growth conditions. (A) Volcano plot of the differentially regulated genes between COG and PLG conditions. For each gene, the rRPKM value (RPKM [COG]/RPKM [PLG]) was plotted against its respective Fisher's exact test P value. P values smaller than $1e^{-290}$ were reported as $1e^{-290}$. The dotted line running parallel to the x-axis indicates the statistical cutoff ($P < 5.5e^{-6}$), whereas the dotted line running parallel to the y-axis indicates the biological cutoff (twofold difference in RPKM between COG and PLG). The red-colored and blue-colored dots correspond to upregulated and downregulated genes between COG and PLG, respectively. (B) Pie chart showing the partitioning of the 9,887 *A. fumigatus* genes with respect to their expression profile in the two growth conditions.

not only are our results unaffected by biological or technical replication issues, but also that they hold across different *A. fumigatus* strains.

Designing and executing a RNA-seq experiment

RNA-seq is a very versatile tool that can be used to address a wide variety of basic and applied science questions, from increasing the genomic depth of the tree of life[10] to characterizing the genetic makeup of cancer in the human body.[41] Consequently, the design of a RNA-seq experiment will vary depending on the question asked and the nature of the investigation, which will in turn determine the acquisition of RNA-seq data, its handling and analysis, as well as the pursuit of follow-up experiments.

In contrast to many other types of bioinformatics analyses where the tools (e.g., the BLAST algorithm[42] is the near universal choice for examining the identity of a sequence) and databases (e.g., the PFAM database[43] is one of a few standard protein domain databases) are well established, the toolkit for the analysis of RNA-seq data is far less well defined, largely because both the technology and its bioinformatics tools are not only very new but also rapidly changing. Although this pace of change

in technology and software makes the design of benchmark studies challenging, a very useful resource for RNA-seq and NGSTs in general is the http://seqanswers.com/ website that aims to provide "an information resource and user-driven community focused on all aspects of next-generation genomics," and which routinely hosts discussions on a variety of topics such as analysis practices and software choice or performance. Although RNA-seq, and NGSTs in general, are touted as cheap technologies, it should be emphasized that the most important challenge to applying this technology is the cost associated with the bioinformatics analysis of the data.

Because of the large-scale nature of RNA-seq experiments, considerable attention is also required for choosing the appropriate experimental design. Fang and Cui[44] recently described a number of experimental design principles that require careful consideration, some of which were discussed in the paragraphs above, including randomization of samples, technical and biological replication of the experiment, the depth and type of sequencing that needs to be performed, as well as whether validation of the results is required by an independent approach, such as qRT-PCR.

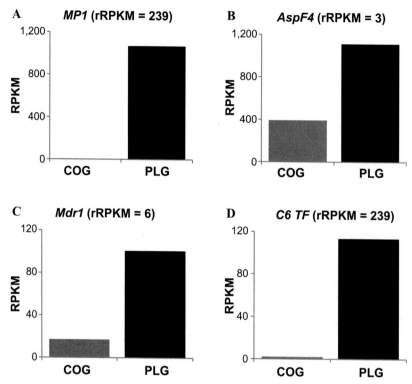

Figure 5. Examples of differentially regulated genes from specific functional categories that constitute candidates for the observed pathobiological and morphological differences between the two conditions. (A) The cell wall galactomannoprotein *MP1*(Afu4g03240). (B) The allergen *AspF4* (Afu2g03830). (C) The ABC multidrug transporter *Mdr1* (Afu5g06070). (D) The C6 transcription factor of the fumitremorgin secondary metabolism gene cluster (Afu8g00420). In each graph, the *x*-axis corresponds to the two growth conditions (COG in black and PLG in gray) and the *y*-axis to the RPKM expression value of the corresponding gene in each of the two conditions. The rRPKM (RPKM [COG]/RPKM [PLG]) value for each comparison is reported in parentheses next to each gene's name.

Conclusions

Even though NGSTs and RNA-seq are less than a decade old, it is abundantly clear that they have dramatically altered the landscape of functional genomics studies in nonmodel organisms. Analysis of the data produced by a single experiment in *A. fumigatus* has uncovered tens of putative unannotated and hundreds of novel small genes, thousands of SNPs, and hundreds of candidates for downstream functional experiments to identify the molecular basis of colony growth and its potential role in the establishment of some forms of aspergillosis. In the near future, we anticipate that RNA-seq applications will not only lead to far greater understanding of the parts, structure, and function of the *A. fumigatus* genome, but also will identify the key differences between *in vitro* and *in vivo* models of the disease, as well as define the molecular interactions between the human host and the fungal pathogen during infection.

Acknowledgments

This work was conducted in part using the resources of the Advanced Computing Center for Research and Education at Vanderbilt University. J.G.G. is funded by the graduate program in biological sciences at Vanderbilt University and the National Institute of Allergy and Infectious Diseases, National Institutes of Health (NIH, NIAID: F31AI091343-01). The content is solely the responsibility of the authors and does not necessarily represent the official views of the NIAID or the NIH. Work in J.-P.L.'s *Aspergillus* lab is partly funded by the ESF Grant Fuminomics and the ALLFUN FP7 project.

Research in A.R.'s lab is supported by the Searle Scholars Program and the National Science Foundation (DEB-0844968).

Conflicts of interest

The authors declare no conflicts of interest.

References

1. Rokas, A. & P. Abbot. 2009. Harnessing genomics for evolutionary insights. *Trends Ecol. Evol.* **24:** 192–200.
2. Gibbons, J.G. *et al.* 2009. Benchmarking next-generation transcriptome sequencing for functional and evolutionary genomics. *Mol. Biol. Evol.* **26:** 2731–2744.
3. Glenn, T.C. 2011. Field guide to next-generation DNA sequencers. *Mol. Ecol. Resour.* **11:** 759–769.
4. Gilad, Y., J.K. Pritchard & K. Thornton. 2009. Characterizing natural variation using next-generation sequencing technologies. *Trends Genet.* **25:** 463–471.
5. Kahvejian, A., J. Quackenbush & J.F. Thompson. 2008. What would you do if you could sequence everything? *Nat. Biotechnol.* **26:** 1125–1133.
6. Wang, Z., M. Gerstein & M. Snyder. 2009. RNA-seq: a revolutionary tool for transcriptomics. *Nat. Rev. Genet.* **10:** 57–63.
7. Wold, B. & R.M. Myers. 2008. Sequence census methods for functional genomics. *Nat. Meth.* **5:** 19–21.
8. Ozsolak, F. & P.M. Milos. 2011. RNA sequencing: advances, challenges and opportunities. *Nat. Rev. Genet.* **12:** 87–98.
9. Garber, M., M.G. Grabherr, M. Guttman & C. Trapnell. 2011. Computational methods for transcriptome annotation and quantification using RNA-seq. *Nat. Meth.* **8:** 469–477.
10. Hittinger, C.T., M. Johnston, J.T. Tossberg & A. Rokas. 2010. Leveraging skewed transcript abundance by RNA-seq to increase the genomic depth of the tree of life. *Proc. Natl. Acad. Sci. U. S. A.* **107:** 1476–1481.
11. Geiser, D.M. *et al.* 2007. The current status of species recognition and identification in *Aspergillus. Stud. Mycol.* **59:** 1–10.
12. Denning, D.W. 1998. Invasive aspergillosis. *Clin. Infect. Dis.* **26:** 781–803.
13. Latge, J.P. 1999. *Aspergillus fumigatus* and aspergillosis. *Clin. Microbiol. Rev.* **12:** 310–350.
14. Morgan, J. *et al.* 2005. Incidence of invasive aspergillosis following hematopoietic stem cell and solid organ transplantation: interim results of a prospective multicenter surveillance program. *Med. Mycol.* **43**(Suppl. 1): S49–S58.
15. Schmitt, H.J., A. Blevins, K. Sobeck & D. Armstrong. 1990. *Aspergillus* species from hospital air and from patients. *Mycoses* **33:** 539–541.
16. Klich, M.A. 2002. Biogeography of *Aspergillus* species in soil and litter. *Mycologia* **94:** 21–27.
17. Shelton, B.G., K.H. Kirkland, W.D. Flanders & G.K. Morris. 2002. Profiles of airborne fungi in buildings and outdoor environments in the United States. *Appl. Environ. Microbiol.* **68:** 1743–1753.
18. Klich, M.A. 2009. Health effects of *Aspergillus* in food and air. *Toxicol. Ind. Health* **25:** 657–667.
19. Raper, K.B. & D.I. Fennell. 1965. *The Genus Aspergillus.* Williams & Wilkins. Baltimore.
20. Loussert, C. *et al.* 2010. *In vivo* biofilm composition of *Aspergillus fumigatus. Cell. Microbiol.* **12:** 405–410.
21. Beauvais, A. *et al.* 2007. An extracellular matrix glues together the aerial-grown hyphae of *Aspergillus fumigatus. Cell. Microbiol.* **9:** 1588–1600.
22. Gibbons, J.G. *et al.* 2012. Global transcriptome changes underlying colony growth in the opportunistic human pathogen *Aspergillus fumigatus. Eukaryot. Cell* **11:** 68–78.
23. Nierman, W.C. *et al.* 2005. Genomic sequence of the pathogenic and allergenic filamentous fungus *Aspergillus fumigatus. Nature* **438:** 1151–1156.
24. Yandell, M. & D. Ence. 2012. A beginner's guide to eukaryotic genome annotation. *Nat. Rev. Genet.* **13:** 329–342.
25. Trapnell, C. *et al.* 2010. Transcript assembly and quantification by RNA-seq reveals unannotated transcripts and isoform switching during cell differentiation. *Nat. Biotechnol.* **28:** 511–515.
26. Guttman, M. *et al.* 2010. *Ab initio* reconstruction of cell type-specific transcriptomes in mouse reveals the conserved multi-exonic structure of lincRNAs. *Nat. Biotechnol.* **28:** 503–510.
27. Trapnell, C., L. Pachter & S.L. Salzberg. 2009. TopHat: discovering splice junctions with RNA-seq. *Bioinformatics* **25:** 1105–1111.
28. Hittinger, C.T. *et al.* 2010. Remarkably ancient balanced polymorphisms in a multi-locus gene network. *Nature* **464:** 54–58.
29. Milgroom, M.G. 1996. Recombination and the multilocus structure of fungal populations. *Annu. Rev. Phytopathol.* **34:** 457–477.
30. Taylor, J.W., E. Turner, J.P. Townsend, *et al.* 2006. Eukaryotic microbes, species recognition and the geographic limits of species: examples from the kingdom Fungi. *Philos. Trans. R. Soc. Lond. B Biol. Sci.* **361:** 1947–1963.
31. Debeaupuis, J.P., J. Sarfati, V. Chazalet & J.P. Latge. 1997. Genetic diversity among clinical and environmental isolates of *Aspergillus fumigatus. Infect. Immun.* **65:** 3080–3085.
32. Pringle, A. *et al.* 2005. Cryptic speciation in the cosmopolitan and clonal human pathogenic fungus *Aspergillus fumigatus. Evolution* **59:** 1886–1899.
33. Rydholm, C., G. Szakacs & F. Lutzoni. 2006. Low genetic variation and no detectable population structure in *Aspergillus fumigatus* compared to closely related *Neosartorya* species. *Eukaryot. Cell* **5:** 650–657.
34. Klaassen, C.H., J.G. Gibbons, N. Fedorova, *et al.* 2012. Evidence for genetic differentiation and variable recombination rates among Dutch populations of the opportunistic human pathogen *Aspergillus fumigatus. Mol. Ecol.* **21:** 57–70.
35. Li, H., J. Ruan & R. Durbin. 2008. Mapping short DNA sequencing reads and calling variants using mapping quality scores. *Genome Res.* **18:** 1851–1858.
36. Klaassen, C.H. 2009. MLST versus microsatellites for typing *Aspergillus fumigatus* isolates. *Med. Mycol.* **47**(Suppl. 1): S27–S33.
37. Bruns, S. *et al.* 2010. Functional genomic profiling of *Aspergillus fumigatus* biofilm reveals enhanced production of the mycotoxin gliotoxin. *Proteomics* **10:** 3097–3107.
38. Marioni, J.C., C.E. Mason, S.M. Mane, *et al.* 2008. RNA-seq: an assessment of technical reproducibility and comparison with gene expression arrays. *Genome Res.* **18:** 1509–1517.

39. Auer, P.L. & R.W. Doerge. 2010. Statistical design and analysis of RNA sequencing data. *Genetics* **185:** 405–416.

40. Bruno, V.M. *et al.* 2010. Comprehensive annotation of the transcriptome of the human fungal pathogen *Candida albicans* using RNA-seq. *Genome Res.* **20:** 1451–1458.

41. Maher, C.A. *et al.* 2009. Transcriptome sequencing to detect gene fusions in cancer. *Nature* **458:** 97–101.

42. Altschul, S.F. *et al.* 1997. Gapped BLAST and PSI-BLAST: a new generation of protein database search programs. *Nucleic Acids Res.* **25:** 3389–3402.

43. Finn, R.D. *et al.* 2006. Pfam: clans, web tools and services. *Nucleic Acids Res.* **34:** D247–D251.

44. Fang, Z. & X. Cui. 2011. Design and validation issues in RNA-seq experiments. *Brief. Bioinform.* **12:** 280–287.

Ann. N.Y. Acad. Sci. ISSN 0077-8923

Conservation in *Aspergillus fumigatus* of pH-signaling seven transmembrane domain and arrestin proteins, and implications for drug discovery

Elaine M. Bignell

Section of Microbiology, Faculty of Medicine, Imperial College London, London, UK

Address for correspondence: Elaine M. Bignell, Section of Microbiology, Flowers Building Level 5, Imperial College London, Exhibition Road, London SW7 2AZ. e.bignell@imperial.ac.uk

Adaptation to extracellular pH is a major challenge to fungal pathogens that infect mammalian hosts. Among pH responses mounted by diverse fungal pathogens there is a high degree of molecular conservation. This, coupled with the absence of such signaling pathways in mammalian cells, suggests that this crucial fungal survival mechanism might provide a useful means of limiting a broad spectrum of infectious fungal growth. PacC/Rim signaling converts extracellular cues, perceived by the fungal cell at extremes of ambient pH, into a cellular signal moderating the activation and/or derepression of multiple pH-sensitive gene functions including enzymes, permeases, and transporters. Signal transduction via the fungal PacC/Rim pathway involves a seven transmembrane domain (7TMD) receptor–arrestin protein complex. This review will discuss, with particular attention to *Aspergillus fumigatus* (the major mold pathogen of humans), the conservation of PacC/Rim signal reception proteins, and protein domains, required for tolerance of pH change, and pathogenicity, and the significance of such molecules as targets for interventive therapies.

Keywords: *Aspergillus*; pH receptor; arrestin; plasma membrane; signal transduction

Introduction

A growing body of transcriptome studies have revealed key stresses imposed by mammalian hosts upon infecting fungal pathogens.[1–5] Among such stresses, those imposed by extremes of pH are relevant to multiple fungal pathogens of man, and have repeatedly been shown to influence the growth of infecting fungi in various host niches including the bloodstream, lung, vagina, cornea, and phagolysosome of innate immune cells.[1–3] Mutations that subvert the, usually robust, tolerance of these organisms to extracellular pH flux invariably result in modulation of virulence, most often severely abrogating infectious growth.[4–8]

A highly conserved fungal signaling pathway, termed PacC/Rim, comprising seven (thus far identified) proteins, safeguards the economy of cellular metabolism under pH stress by restricting the expression of pH-sensitive gene products to conditions at which they are (i) required by the cell and (ii) optimally functional.[1] The PacC/Rim signaling pathway has been most well studied in *Aspergillus nidulans*, *Saccharomyces cerevisiae* and *Candida albicans*, in which the key effectors of pH adaptation in all three species are homologous zinc finger DNA-binding proteins called PacC (in *Aspergillus* species) or Rim101 (in yeasts).[1]

The fungal PacC/Rim pH signaling mechanism represents a unique regulatory paradigm for eukaryotic cells as it appears to have coopted components of the fungal multivesicular body (MVB) sorting machinery to effect posttranslational processing of the PacC/Rim101 transcription factor. Most of the components of the ESCRT (endosomal sorting complex required for transport) machinery involved in the MVB (multivesicular body) sorting pathway,[9,10] play an essential role in the PacC/Rim signaling pathway.[1,11–16]

doi: 10.1111/j.1749-6632.2012.06814.x

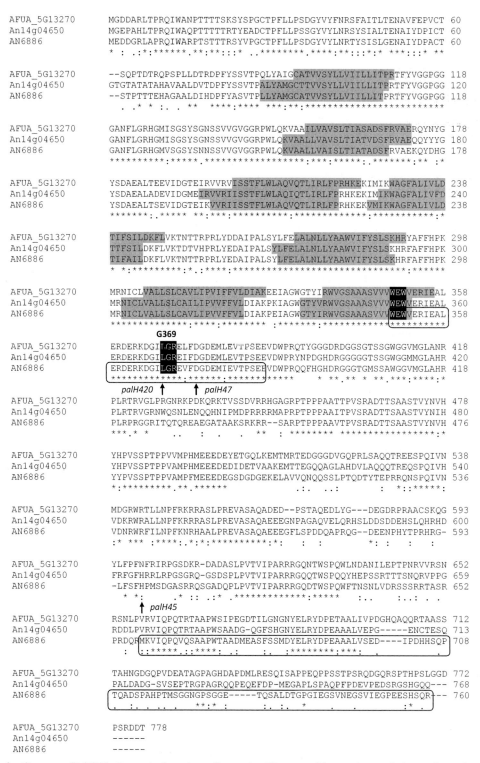

Figure 1. Alignment of PalH/Rim21 proteins from *Aspergillus* species. Alignment of the putative translation products of open reading frames AFUA_5G13270 (*A. fumigatus*), An14g04650 (*A. niger*) and AN6886 (*A. nidulans*). Polypeptide sequences were obtained from the AspGD data base (http://www.aspgd.org/) and aligned using ClustalW2 (http://www.ebi.ac.uk/Tools/msa/clustalw2/).

Signal reception likely involves a GPCR-like seven transmembrane domain (7TMD) pH sensor, PalH/Rim21p, and a fungal arrestin-like protein, PalF/Rim8p. Both proteins lack conserved homologs in mammalian cells and PalF/Rim8p has been implicated by bioinformatic analyses as a potential target for antifungal drugs, being essential for virulence in mammalian hosts and restricted to fungal cells.[17]

The interactions between 7TMD proteins and their cognate arrestins represent the most common class of target for existing pharmaceutical drugs.[18] Accordingly, PalH/Rim21 and PalF/Rim8 present putative targets for development of novel antifungal therapies; this review will focus upon the role of these proteins, and their interacting partners in fungal pathogens, the conservation of these crucial pH signaling proteins in *A. fumigatus*, and the mechanistic features of this protein partnership that might lend themselves to the furtherance of drug discovery.

PalH and PalI proteins: putative pH sensors in *Aspergillus* species

Evidence for participation of two integral membrane proteins in *A. nidulans* PacC-mediated pH signaling was obtained when cDNAs rescuing the alkaline-sensitive phenotypes of two, independently segregating, mutants, were identified, sequenced and designated as *PalH* and *PalI*.[19] The derived amino acid sequence for the *A. nidulans* PalH protein is predicted to contain 760 amino acids, the amino portion of which is predicted to contain seven hydrophobic transmembrane (TM) domains.[20] In alignment with *A. fumigatus* and *A. niger* translation products, a very high degree of amino acid conservation is evident in regions of predicted transmembrane helices (Fig. 1). Preferred topological predictions for *A. fumigatus*, *A. nidulans,* and *A. niger* proteins place the PalH N-termini in the periplasm and the C-termini in the cytoplasm

(Fig. 2). This topology, in the case of *A. fumigatus*, has been experimentally validated by analysis of recombinant proteins, expressed as split-ubiquitin baits in *S. cerevisiae*.[21] Mutational evidence indicates the importance of the PalH 7TM segment-containing N-terminal moiety, because *A. nidulans* mutants lacking the C-terminal 383 residues (encompassing nearly all of the long hydrophilic C-terminus) retain residual PalH function.[20]

In *A. nidulans*, co-overexpression of PalH–GFP with either a PalI–GFP or a PalI–(HA)$_3$ fusion protein leads exclusively to a plasma membrane localization of PalH–GFP and indicates that PalI is required in stoichiometrically similar quantities to PalH to assist with appropriate subcellular localization of the PalH pH sensor.[22] This observation, and the ability of PalI-null mutants to tolerate alkaline pH (albeit to a much lesser extent than the wild type), led to the proposal that PalI acts to assist the plasma membrane localization of PalH.[22] A schematic summary of the proteins discussed in this review, and their roles in fungal pH signal reception and transduction, is provided in Figure 4.

PalH homologues in *Yarrowia lipolytica*, *S. cerevisiae*, and *C. albicans*

A PalH homolog from the ascomycetous yeast *Yarrowia lipolytica* was cloned by complementation and was found to have similarity to an (at the time) unidentified *S. cerevisiae* open reading frame YNL294c, now named *RIM21*.[23] A second *S. cerevisiae* pH sensor, *sc*Dfg16p, was discovered through a gene deletion mutant screen, using a synthetic plasmid-borne Rim101p promoter-lacZ fusion to identify mutants with a deficiency of *Sc*Rim101p-dependent repression.[24] Similar to a *Sc*Rim21p-null mutant, isolates lacking the *Sc*Dfg16p protein fail to process the *Sc*Rim101p transcription factor and are defective in haploid invasive growth.[24] The *DFG16* gene is conserved in *C. albicans* in which a *dfg16*Δ/*dfg16*Δ deletion mutant is defective

---

Transmembrane helices, predicted using the TMPRED server (http://www.ch.embnet.org/software/TMPRED_form.html) are shaded in gray. PalF/Rim8 arrestin-binding domains, as predicted by screening of *A. nidulans* homologues in yeast two hybrid analyses,[27] are indicated by unshaded gray rectangles). Residues highlighted with black shading occur within a PalF binding domain and are conserved among yeasts and Aspergilli.[27] The highly conserved glycine 369 residue, which is essential for PalF-PalH binding *in vitro* and *in vivo*,[27] is indicated above the alignment as G369. The site of an *A. nidulans palH* loss-of-function mutation, which truncates the protein after residue 368 and removes glycine 369, is indicated by an arrow. The sites of two *A. nidulans palH* partial loss-of-function mutations, *palH47* and *palH45* are also indicated.

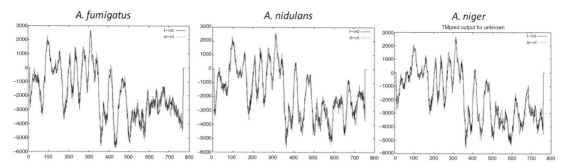

Figure 2. Hydropathy plots for *Aspergillus palH* gene products. Transmembrane helices were predicted using the TMPRED server (http://www.ch.embnet.org/software/TMPRED_form.html). The software plots the hydropathy index of the predicted translation product (*y*-axis) as a function of its location in the polypeptide chain (*x*-axis).

for alkaline-induced filamentation and virulence in mammalian hosts.[4]

Arrestin-mediated pH signaling in *Aspergillus* species

Arrestins modulate intracellular signaling through interactions with cytoplasmic domains of 7TMD receptors. Often engaging the carboxy terminus and several cytoplasmic loops of the receptor, arrestins have been documented as both positive and negative effectors of signal transduction acting, variously, to shut off G-protein-mediated signaling, target 7TMD receptors for endocytic internalization and redirect GPCR signaling to a variety of G-protein-independent pathways. Arrestins were originally named after their functional propensity for effecting receptor desensitization through steric hindrance of G protein coupling to activated 7TMD receptors. Recently, however, arrestins are becoming much more widely recognized as versatile adaptor molecules that are also capable of activating protein signaling cascades independently of G protein function.[25]

The alkaline-sensitive phenotype of a *palF15* *A. nidulans* mutant, which was originally isolated and characterized as aberrantly expressing acid and alkaline phosphatases, suggested a role for PalF in PacC-mediated pH signaling. Cloning, by reversal of alkaline sensitivity, and sequencing of the *palF* gene yielded little initial insight as to the function of the protein.[26] Subsequent efforts to identify proteins binding to *A. nidulans* PalF, by using yeast two hybrid analyses with a PalF–LexA fusion protein and a library of *A. nidulans* protein preys, identified PalH C-terminal polypeptides as strong binders.[27] This clue to PalF function, coupled with further

scrutiny of the putative PalF translation product sequence, was strongly suggestive of an arrestin-like function for PalF/Rim8. *A. nidulans* PalF was found to contain both N- (residues 64–206) and C-terminal (residues 306–463) arrestin domains and two hybrid analyses identified strong interactions with two different regions of the *A. nidulans* PalH C-terminus.[27] Investigation of the binding requirements for *A. nidulans* PalF-PalH interaction employed yeast two hybrid mapping and identified the N-terminal arrestin domain, plus 63 further N-terminal amino acid residues as being sufficient for PalH interaction. The N-terminal arrestin domain is highly conserved among *Aspergillus* species (Fig. 3), including *A. fumigatus*, and yeasts.[27] Lack of the *A. nidulans* highly conserved glycine 369 residue and substitution of glycine 369 in yeast two hybrid PalH baits leads, respectively, to a loss of function phenotype *in vivo* and loss of PalF interaction *in vitro*.[27] These findings portray a crucial role for the interaction between receptor and arrestin for PacC/Rim signaling in *Aspergillus* species. In support of this protein partnership as a target for anti-*Aspergillus* therapy, a yeast membrane two hybrid strategy, applied to test the interaction between full length *A. fumigatus* PalH and PalF homologs in living fungal cells, also reveals a potent interaction.[21] Moreover, PacC- and PalH-null mutants of *A. fumigatus* are severely attenuated for virulence in murine models of pulmonary aspergillosis (Bertuzzi and Bignell *et al.*, in preparation).

Two independent regions of the *A. nidulans* PalH cytoplasmic C-terminal moiety (residues 349–385 and 657–760) were identified as PalF binders (indicated by gray rectangles in Fig. 1) by two hybrid analyses.[27] The physiological significance of this

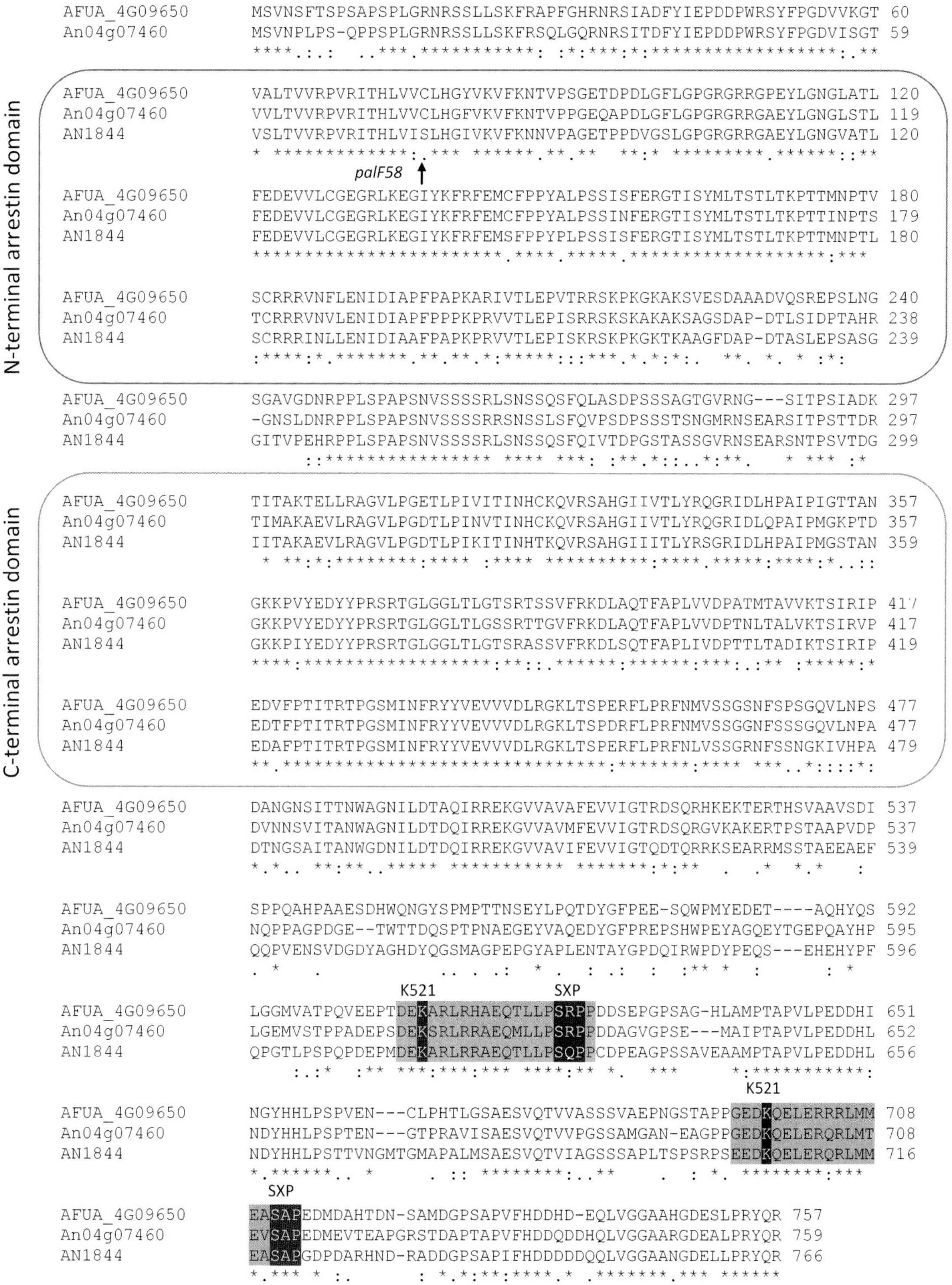

N-terminal arrestin domain

```
AFUA_4G09650    MSVNSFTSPSAPSPLGRNRSSLLSKFRAPFGHRNRSIADFYIEPDDPWRSYFPGDVVKGT  60
An04g07460      MSVNPLPS-QPPSPLGRNRSSLLSKFRSQLGQRNRSITDFYIEPDDPWRSYFPGDVISGT  59
                ****.::.  ..***.***********: :*:***:*:*************** :** **

AFUA_4G09650    VALTVVRPVRITHLVVCLHGYVKVFKNTVPSGETDPDLGFLGPGRGRRGPEYLGNGLATL  120
An04g07460      VVLTVVRPVRITHLVVCLHGFVKVFKNTVPPGEQAPDLGFLGPGRGRRGAEYLGNGLSTL  119
AN1844          VSLTVVRPVRITHLVISLHGIVKVFKNNVPAGETPPDVGSLGPGRGRRGAEYLGNGVATL  120
                * *************.:.*** ******.**.** **:* ********* .*****.:.**
                                   palF58 ↑
AFUA_4G09650    FEDEVVLCGEGRLKEGIYKFRFEMCFPPYALPSSISFERGTISYMLTSTLTKPTTMNPTV  180
An04g07460      FEDEVVLCGEGRLKEGIYKFRFEMCFPPYALPSSINFERGTISYMLTSTLTKPTTINPTS  179
AN1844          FEDEVVLCGEGRLKEGIYKFRFEMSFPPYPLPSSISFERGTISYMLTSTLTKPTTMNPTL  180
                ************************ ****.*****.**********.************:***

AFUA_4G09650    SCRRRVNFLENIDIAPFPAPKARIVTLEPVTRRSKPKGKAKSVESDAAADVQSREPSLNG  240
An04g07460      TCRRRVNVLENIDIAPFPPPKPRVVTLEPISRRSKSKAKAKSAGSDAP-DTLSIDPTAHR  238
AN1844          SCRRRINLLENIDIAAFPAPKPRVVTLEPISKRSKPKGKTKAAGFDAP-DTASLEPSASG  239
                :****:*.*******.**.**.**.*****::***.*.*:*:.   **.*. * :*:
```

```
AFUA_4G09650    SGAVGDNRPPLSPAPSNVSSSSRLSNSSQSFQLASDPSSSAGTGVRNG---SITPSIADK  297
An04g07460      -GNSLDNRPPLSPAPSNVSSSSRRSNSSLSFQVPSDPSSSTSNGMRNSEARSITPSTTDR  297
AN1844          GITVPEHRPPLSPAPSNVSSSSRLSNSSQSFQIVTDPGSTASSGVRNSEARSNTPSVTDG  299
                :::*************** **** ***: :**.*::..*:**.    * *** :*
```

C-terminal arrestin domain

```
AFUA_4G09650    TITAKTELLRAGVLPGETLPIVITINHCKQVRSAHGIIVTLYRQGRIDLHPAIPIGTTAN  357
An04g07460      TIMAKAEVLRAGVLPGDTLPINVTINHCKQVRSAHGIIVTLYRQGRIDLQPAIPMGKPTD  357
AN1844          IITAKAEVLRAGVLPGDTLPIKITINHTKQVRSAHGIIITLYRSGRIDLHPAIPMGSTAN  359
                 * **:*:********:**** :**** ***********:****.*****.****:..::

AFUA_4G09650    GKKPVYEDYYPRSRTGLGGLTLGTSRTSSVFRKDLAQTFAPLVVDPATMTAVVKTSIRIP  417
An04g07460      GKKPVYEDYYPRSRTGLGGLTLGSSRTTGVFRKDLAQTFAPLVVDPTNLTALVKTSIRVP  417
AN1844          GKKPIYEDYYPRSRTGLGGLTLGTSRASSVFRKDLSQTFAPLIVDPTTLTADIKTSIRIP  419
                ****:******************:**:..******:******:.***  :** .*****:*

AFUA_4G09650    EDVFPTITRTPGSMINFRYYVEVVVDLRGKLTSPERFLPRFNMVSSGSNFSPSGQVLNPS  477
An04g07460      EDTFPTITRTPGSMINFRYYVEVVVDLRGKLTSPDRFLPRFNMVSSGGNFSSSGQVLNPA  477
AN1844          EDAFPTITRTPGSMINFRYYVEVVVDLRGKLTSPERFLPRFNLVSSGRNFSSNGKIVHPA  479
                **.*****************************.*:*******::***** ***.:**.:::*
```

```
AFUA_4G09650    DANGNSITTNWAGNILDTAQIRREKGVVAVAFEVVIGTRDSQRHKEKTERTHSVAAVSDI  537
An04g07460      DVNNSVITANWAGNILDTDQIRREKGVVAVMFEVVIGTRDSQRGVKAKERTPSTAAPVDP  537
AN1844          DTNGSAITANWGDNILDTDQIRREKGVVAVIFEVVIGTQDTQRRKSEARRMSSTAEEAEF  539
                *.*..  **:**.***** *:************ *:*******:*:**  . .* *.* :

AFUA_4G09650    SPPQAHPAAESDHWQNGYSPMPTTNSEYLPQTDYGFPEE-SQWPMYEDET----AQHYQS  592
An04g07460      NQPPAGPDGE--TWTTDQSPTPNAEGEYVAQEDYGFPREPSHWPEYAGQEYTGEPQAYHP  595
AN1844          QQPVENSVDGDYAGHDYQGSMAGPEPGYAPLENTAYGPDQIRWPDYPEQS---EHEHYPF  596
                . *   .    ..  :: .* .  .: : :** *   :   :  * * *
                                            K521           SXP
AFUA_4G09650    LGGMVATPQVEEPTDEKARLRHAEQTLLPSRPPDDSEPGPSAG-HLAMPTAPVLPEDDHI  651
An04g07460      LGEMVSTPPADEPSDEKSRLRRAEQMLLPSRPPDDAGVGPSE---MAIPTAPVLPEDDHL  652
AN1844          QPGTLPSPQPDEPMDEKARLRRAEQTLLPSQPPCDPEAGPSSAVEAAMPTAPVLPEDDHL  656
                :.:*  :** :** :**:*:**:****  *** * .   ***   *.***********:
                                                               K521
AFUA_4G09650    NGYHHLPSPVEN---CLPHTLGSAESVQTVVASSSVAEPNGSTAPPGEDKQELERRRLMM  708
An04g07460      NDYHHLPSPTEN---GTPRAVISAESVQTVVPGSSAMGAN-EAGPPGEDKQELERQRLMT  708
AN1844          NDYHHLPSTTVNGMTGMAPALMSAESVQTVIAGSSSAPLTSPSRPSEEDKQELERQRLMM  716
                *.*******..  *     . :: *******. **.**  .   ********:***
                SXP
AFUA_4G09650    EASAPEDMDAHTDN-SAMDGPSAPVFHDDHD-EQLVGGAAHGDESLPRYQR  757
An04g07460      EVSAPEDMEVTEAPGRSTDAPTAPVFHDDQDDHQLVGGAARGDEALPRYQR  759
AN1844          EASAPGDPDARHND-RADDGPSAPIFHDDDDDQQLVGGAANGDELLPRYQR  766
                *.***.*  :    . ..*.*:**:****. * ******** ***. ***********
```

Figure 3. Alignment of PalF/Rim8 proteins from *Aspergillus* species. Alignment of the putative translation products of open reading frames AFUA_4G09560 (*A. fumigatus*, An04g07460 (*A. niger*) and AN1844 (*A. nidulans*). Polypeptide sequences were obtained from the AspGD data base (http://www.aspgd.org/) and aligned using ClustalW2 (http://www.ebi.ac.uk/Tools/msa/clustalw2/). N- and

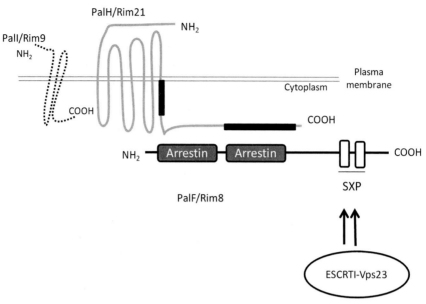

Figure 4. Schematic representation of pH-sensing receptor and arrestin proteins in *Aspergillus* species. Two putative pH sensors, PalH/Rim21 and PalI/Rim9 are located in the plasma membrane of *Aspergillus* species.[22,29] PalH/Rim21 proteins are predicted to contain seven transmembrane helices, while PalI/Rim9 proteins (dotted line) that are dispensable for full alkaline tolerance are predicted to contain three transmembrane helices. Black shaded rectangles on PalH/Rim21 represent PalF/Rim8-binding regions, as predicted by yeast two hybrid analyses.[27] Gray shaded arrestin boxes on PalF/Rim9 represent N- and C-terminal conserved arrestin domains and white boxes represent the sites of SXP motifs and conserved lysine residues, ubiquitination of which is likely, based upon studies in *S. cervisiae*, to lead to ESCRT-I Vps23 recruitment, and which is essential for pH signaling in *Aspergillus* species.

interaction is likely to extend, at least, to promotion of the PalH-dependent phosphorylation and ubiquitination of native *A. nidulans* PalF protein, which occurs upon acid to alkaline shift.[27] The working model for alkaline-induced PacC signaling therefore involves signal- and 7TMD receptor–dependent arrestin phosphorylation and ubiquitination, the latter of which two modifications has been demonstrated, via forced expression of a PalF-ubiquitin translational fusion, to be essential for PacC/Rim signaling.[28] Thus, in *A. nidulans*, the 7TMD PalH pH receptor modulates accessibility of PalF, in an as yet uncharacterized but alkaline-dependent manner, to a ubiquitinating entity. The *A. nidulans* ESCRT-1 component Vps23 interacts with PalF,[29] and in keeping with a role for Vps23 in directing the alkaline-dependent recruitment of Rim signal-

ing pathway members, coimmunoprecipitates exclusively ubiquitinated forms of PalF.[29]

In mammalian GPCR-arrestin signaling β-arrestin ubiquitination promotes endocytic internalization of arrestin–receptor complexes. Based upon this knowledge, the involvement of ESCRT components in PacC/Rim signaling has long been postulated to facilitate endocytosis of a signaling complex, connecting alkaline signal reception with downstream Pac/Rim transcription factor processing.[1] The most recent studies of *A. nidulans* pH signaling, by harnessing the power of dual channel epifluorescence microscopy, have altered this view by demonstrating the sequential recruitment of multiple Pal signaling components to cortical regions of the cell. Vps23 and the ESCRT-III, Vps32 interactors, PalA, and PalC transiently colocalize at

C-terminal arrestin domains, predicted by comparison to the study of Herranz *et. al.*,[27] are indicated by gray, unshaded rectangles. The site of an *A. nidulans*, loss-of-function missense mutation, *palF58*, is indicated. Conserved sequence boxes containing SXP motifs, which direct Vps23 interactions,[15] are shaded in gray and within these, the SXP motifs and highly conserved lysines (thought to be candidates for ubiquitination) are indicated by black shading and, respectively, the terms K521 and SXP.

alkaline ambient pH to cortical structures at the plasma membrane of the cell.[29] These new findings provide robust support for a model in which pH signaling in *Aspergillus* species takes place in plasma membrane-associated protein complexes.

Arrestin-mediated pH signaling in non-*Aspergillus* fungal species

In *Aspergillus* species and ascomycetous yeasts, PacC/Rim transcription factor processing requires multiple components of the cellular endosomal sorting complexes required for trafficking (ESCRT) machinery that facilitates the trafficking of ubiquitylated proteins from endosomes to lysosomes via MVBs.[9] In *S. cerevisiae* the PalF homolog, *Sc*Rim8, binds to the carboxy terminus of at least one of the two PalH homologs, *Sc*Rim21, as well as to the ESCRT-1 subunit *Sc*Vps23.[15] Vps23 interaction is mediated, in *S. cerevisiae*, by an SXP sequence motif, which is conserved at the carboxy termini of Rim8 homologs from yeasts to *Aspergillus* species (Fig. 3), and occurs as a distinct single copy variant in *S. cerevisiae* and *C. albicans*[15] but is doubly present in the C-termini of *Aspergillus* PalF translation products (Fig. 3). In *S. cerevisiae*, alanine substitutions in the SXP motif abolish Vps23 interactions *in vitro* without affecting receptor–arrestin interaction.[15] Fungal Rim8/PalF SXP boxes contain, where studied, conserved lysine residues that are candidates for ubiquitination (Fig. 3). Immunoblot analyses in *S. cerevisae* validated the importance of the residue as a site of Rim8 ubiquitination, however, unlike PalF of *A. nidulans*, *Sc*Rim8 ubiquitination occurs independently of pH signaling, as evidenced by omnipresence of the ubiquitinylated species at both high and low pHs.[15] A further distinction between *S. cerevisiae* and *A. nidulans* Rim8/PalF ubiquitination is their differential dependence upon pH and Rim/Pal pH signaling, whereby Rim8 is ubiquitinated independently of Rim signaling in a non-pH-dependent manner, while *A. nidulans* PalF ubiquitination occurs in a PalH receptor– and pH-dependent manner. *Sc*Rim8 ubiquitination, which does not occur in a *vps23* mutant, is not important for Rim signaling in *S. cerevisiae*. A NEDD4 E3 ubiquitin (Ub) ligase, Rsp5, was found to direct *Sc*Rim8 ubiquitination, and the postulated role of this protein modification, namely recruitment of the ESCRT machinery via Vps23, was proven by fusion of Vps23 to C-terminally truncated Rim8 lacking ubiquitination

site and Vps23 interaction site.[15] Translational fusion of Rim8 and Vps23 effectively bypasses the requirement for active Vps23 recruitment and was found to fully restore pH-dependent regulation of Rim101 processing.

In *C. albicans* Rim8 modification is correlated with Rim101 activation, and Rim8 protein levels are reduced concomitantly with processing of Rim101.[30] However, contrary to both *S. cerevisiae* and *A. nidulans* arrestin homologs, *C. albicans* Rim8 is apparently not ubiquitinated but hyperphosphorylated, and this occurs in response to neutral-alkaline pH.[30] Specific Rim8 phosphorylation states are associated with alkaline-induced activation of the pH-sensing pathway such that *Ca*Rim8 hyperphosphorylation occurs during alkalinization of the growth medium. Analysis of *Ca*Rim101 processing in *Ca*Rim mutants revealed that hyperphosphorylation is dependent upon intact Rim101 signaling; moreover, *Ca*Rim8 protein levels decline acutely as pH rises suggesting the presence of a negative feedback regulatory loop.[30] Further immunoblotting and immunoprecipitation analyses identified pH-dependent phosphorylation of the *Ca*Rim21 7TMD receptor, and the novel findings that (a) *Ca*Rim8 complexes with *Ca*Rim101 and (b) *Ca*Rim8 is transported to the vacuole of the cell at alkaline pH where it is likely to be degraded.[30] In the absence of the AAA-ATPase *Ca*Vps4, which is impaired for vacuolar fusion, Rim8 transportation to the vacuole was not observed, thereby explaining constitutive activation of Rim101 signaling and reinforcing the relevance of Rim8-mediated negative feedback in moderating the activity of the signaling pathway. In addition to down regulation via vacuolar delivery and degradation, *Ca*Rim8 has also been found to be transcriptionally repressed in a Rim101-dependent manner.[31,32]

Prevention of PacC/Rim signaling as a therapeutic strategy

As a result of the detailed molecular characterization performed in a range of fungal species, multiple foci for targeted inhibition of fungal pH signaling, as a therapeutic strategy, present themselves. Although originally thought to assemble upon endosomal membranes, the pH "signalosome" is more likely, based upon recent evidence, to be located at cortical structures, having important implications for delivery of inhibitory entities. Abolition of PalH/Rim21–PalF/Rim8 7TMD receptor–arrestin interaction, achievable mutationally via single amino acid substitutions,[27] abrogates pH signaling. The targeted inhibition of this (likely constitutive) protein interaction is a promising strategy, as is inhibition of the structural and/or posttranslational modifications of the receptor–arrestin complex that are required at alkaline pH to promote pH signaling.

Alkaline ambient pH triggers a negative transcriptional feedback mechanism that acts through the *palF* promoter and PalF orthologues, *Sc*Rim8, and *Ca*Rim8, which are targets of negative feedback inhibition in a Rim101-dependent manner. Thus, the PalF/Rim8 arrestin is, from multiple perspectives, a limiting factor for alkaline signaling. The appeal of PalF/Rim8 proteins, and cognate 7TMDs, as a therapeutic target extends even further beyond the simple, but nonredundant, requirement for the proteins in mediating a crucial pathogen survival mechanism in mammalian hosts, and beyond the exquisite detail in which its role in signal transduction is becoming characterized. Recently, comparative genomics, assessing eight fungal pathogen genomes, identified PalF/Rim8 as one of only 10 genes required for survival in the host and absent in the human genome.[17]

Acknowledgments

I am immensely grateful to Herbert N. Arst Jr., Joan Tilburn, and Darius Armstrong James for critical reading of the manuscript. Relevant research in the Bignell laboratory has been supported by the Medical Research Council [G0501164].

Conflicts of interest

The author declares no conflicts of interest.

References

1. Penalva, M.A., J. Tilburn, E. Bignell & H.N. Arst, Jr. 2008. Ambient pH gene regulation in fungi: making connections. *Trends Microbiol.* **16:** 291–300.

2. Davis, D.A. 2009. How human pathogenic fungi sense and adapt to pH: the link to virulence. *Curr. Opin. Microbiol.* **12:** 365–370.

3. De, B.F., F.A. Muhlschlegel, A. Cassone & W.A. Fonzi. 1998. The pH of the host niche controls gene expression in and virulence of Candida albicans. *Infect. Immun.* **66:** 3317–3325.

4. Thewes, S., M. Kretschmar, H. Park, *et al.* 2007. In vivo and ex vivo comparative transcriptional profiling of invasive and non-invasive Candida albicans isolates identifies genes

5. Yuan, X., B.M. Mitchell, X. Hua, *et al.* 2010. The RIM101 signal transduction pathway regulates Candida albicans virulence during experimental keratomycosis. *Invest Ophthalmol. Vis. Sci.* **51:** 4668–4676.

6. Bignell, E., S. Negrete-Urtasun, A.M. Calcagno, *et al.* 2005. The Aspergillus pH-responsive transcription factor PacC regulates virulence. *Mol. Microbiol.* **55:** 1072–1084.

7. O'Meara, T.R., D. Norton, M.S. Price, *et al.* 2010. Interaction of Cryptococcus neoformans Rim101 and protein kinase A regulates capsule. *PLoS Pathog.* **6:** e1000776.

8. Davis, D., J.E. Edwards, Jr., A.P. Mitchell & A.S. Ibrahim. 2000. Candida albicans RIM101 pH response pathway is required for host-pathogen interactions. *Infect. Immun.* **68:** 5953–5959.

9. Hurley, J.H. 2008. ESCRT complexes and the biogenesis of multivesicular bodies. *Curr. Opin. Cell Biol.* **20:** 4–11.

10. Williams, R.L. & S. Urbe. 2007. The emerging shape of the ESCRT machinery. *Nat. Rev. Mol. Cell Biol.* **8:** 355–368.

11. Galindo, A., A. Hervas-Aguilar, O. Rodriguez-Galan, *et al.* 2007. PalC, one of two Bro1 domain proteins in the fungal pH signaling pathway, localizes to cortical structures and binds Vps32. *Traffic* **8:** 1346–1364.

12. Cornet, M., F. Bidard, P. Schwarz, *et al.* 2005. Deletions of endocytic components VPS28 and VPS32 affect growth at alkaline pH and virulence through both RIM101-dependent and RIM101-independent pathways in Candida albicans. *Infect. Immun.* **73:** 7977–7987.

13. Blanchin-Roland, S., C.G. Da & C. Gaillardin. 2005. ESCRT-I components of the endocytic machinery are required for Rim101-dependent ambient pH regulation in the yeast Yarrowia lipolytica. *Microbiology* **151:** 3627–3637.

14. Kullas, A.L., M. Li & D.A. Davis. 2004. Snf7p, a component of the ESCRT-III protein complex, is an upstream member of the RIM101 pathway in Candida albicans. *Eukaryot. Cell* **3:** 1609–1618.

15. Herrador, A., S. Herranz, D. Lara & O. Vincent. 2010. Recruitment of the ESCRT machinery to a putative seven-transmembrane-domain receptor is mediated by an arrestin-related protein. *Mol. Cell Biol.* **30:** 897–907.

16. Rodriguez-Galan, O., A. Galindo, A. Hwervas-Aguilar, *et al.* 2009. Physiological involvement in pH signaling of Vps24-mediated recruitment of Aspergillus PalB cysteine protease to ESCRT-III. *J. Biol. Chem.* **284:** 4404–4412.

17. Abadio, A.K., E.S. Kioshima, M.M. Teixeira, *et al.* 2011. Comparative genomics allowed the identification of drug targets against human fungal pathogens. *BMC Genomics* **12:** 75.

18. Zhang, R. & X. Xie. 2012. Tools for GPCR drug discovery. *Acta Pharmacol. Sin.* **33:** 372–384.

19. Arst, H.N., Jr., E. Bignell & J. Tilburn. 1994. Two new genes involved in signaling ambient pH in Aspergillus nidulans. *Mol. Gen. Genet.* **245:** 787–790.

20. Negrete-Urtasun, S., W. Reiter, E. Diez, *et al.* 1999. Ambient pH signal transduction in Aspergillus: completion of gene characterization. *Mol. Microbiol.* **33:** 994–1003.

21. Bertuzzi, M. & E.M. Bignell. 2012. Sensory perception in fungal pathogens: applications of the split-ubiquitin membrane yeast two hybrid system. *Fungal Biol. Rev.* **25:** 165–171.

22. Calcagno-Pizarelli, A.M., S. Negrete-Urtasun, S.H. Denison, *et al.* 2007. Establishment of the ambient pH signaling complex in Aspergillus nidulans: PalI assists plasma membrane localization of PalH. *Eukaryot. Cell* **6:** 2365–2375.

23. Lambert, M., S. Blanchin-Roland, L.F. Le, *et al.* 1997. Genetic analysis of regulatory mutants affecting synthesis of extracellular proteinases in the yeast Yarrowia lipolytica: identification of a RIM101/pacC homolog. *Mol. Cell Biol.* **17:** 3966–3976.

24. Barwell, K.J., J.H. Boysen, W. Xu & A.P. Mitchell. 2005. Relationship of DFG16 to the Rim101p pH response pathway in Saccharomyces cerevisiae and Candida albicans. *Eukaryot. Cell* **4:** 890–899.

25. Dewire, S.M., S. Ahn, R.J. Lefkowitz & S.K. Shenoy. 2007. Beta-arrestins and cell signaling. *Annu. Rev. Physiol.* **69:** 483–510.

26. MacCheroni, W., Jr., G.S. May, N.M. Martinez-Rossi & A. Rossi. 1997. The sequence of palF, an environmental pH response gene in Aspergillus nidulans. *Gene* **194:** 163–167.

27. Herranz, S., J.M. Rodriguez, H.J. Bussink, *et al.* 2005. Arrestin-related proteins mediate pH signaling in fungi. *Proc. Natl. Acad. Sci. USA* **102:** 12141–12146.

28. Hervas-Aguilar, A., A. Galindo & M.A. Penalva. 2010. Receptor-independent Ambient pH signaling by ubiquitin attachment to fungal arrestin-like PalF. *J. Biol. Chem.* **285:** 18095–18102.

29. Galindo, A., A.M. Calcagno-Pizarelli, H.N. Arst, Jr. & M.A. Penalva. 2012. An ordered pathway for the assembly of fungal ESCRT-containing ambient pH signaling complexes at the plasma membrane. *J. Cell Sci.* **125:** 1784–1795.

30. Gomez-Raja, J. & D.A. Davis. 2012. The beta-arrestin-like protein Rim8 Is hyperphosphorylated and complexes with Rim21 and Rim101 to promote adaptation to neutral-alkaline pH. *Eukaryot. Cell* **11:** 683–693.

31. Porta, A., A.M. Ramon & W.A. Fonzi. 1999. PRR1, a homolog of Aspergillus nidulans palF, controls pH-dependent gene expression and filamentation in Candida albicans. *J. Bacteriol.* **181:** 7516–7523.

32. Bensen, E.S., S.J. Martin, M. Li, *et al.* 2004. Transcriptional profiling in Candida albicans reveals new adaptive responses to extracellular pH and functions for Rim101p. *Mol. Microbiol.* **54:** 1335–1351.

Ann. N.Y. Acad. Sci. ISSN 0077-8923

ANNALS OF THE NEW YORK ACADEMY OF SCIENCES
Issue: *Advances Against Aspergillosis*

Protein targets for broad-spectrum mycosis vaccines: quantitative proteomic analysis of *Aspergillus* and *Coccidioides* and comparisons with other fungal pathogens

Jackson Champer,[1,2] Diana Diaz-Arevalo,[1] Miriam Champer,[1] Teresa B. Hong,[1] Mayyen Wong,[1] Molly Shannahoff,[1,3] James I. Ito,[4] Karl V. Clemons,[5,6,7] David A. Stevens,[5,6,7] and Markus Kalkum[1,2,3]

[1]Department of Immunology, [2]Irell and Manella Graduate School of Biological Sciences, and [3]Eugene and Ruth Roberts Summer Academy, the Beckman Research Institute of the City of Hope, Duarte, California. [4]Division of Infectious Diseases, City of Hope National Medical Center, Duarte, California. [5]California Institute for Medical Research, San Jose, California. [6]Department of Medicine, Santa Clara Valley Medical Center, San Jose, California. [7]Department of Medicine, Stanford University, Stanford, California.

Address for correspondence: Markus Kalkum, Department of Immunology, City of Hope Beckman Research Institute, 1500 East Duarte Road, Duarte, CA 91010, USA. mkalkum@coh.org

Aspergillus species are responsible for most cases of fatal mold infections in immunocompromised patients, particularly in those receiving hematopoietic stem cell transplants. Experimental vaccines in mouse models have demonstrated a promising avenue of approach for the prevention of aspergillosis, as well as infections caused by other fungal pathogens, such as *Coccidioides*, the etiological agent of valley fever (coccidioidomycosis). Here, we investigated the hyphal proteomes of *Aspergillus fumigatus* and *Coccidioides posadasii* via quantitative MS^E mass spectrometry with the objective of developing a vaccine that cross-protects against these and other species of fungi. Several homologous proteins with highly conserved sequences were identified and quantified in *A. fumigatus* and *C. posadasii*. Many abundant proteins from the cell wall of *A. fumigatus* present themselves as possible cross-protective vaccine candidates, due to the high degree of sequence homology to other medically relevant fungal proteins and low homologies to human or murine proteins.

Keywords: quantitative proteomics; mass spectrometry; panfungal vaccine; *Aspergillus*; *Coccidioides*

Introduction

Pathogenic fungi, particularly the mold *Aspergillus fumigatus*, cause thousands of deaths per year,[1] mostly among immunosuppressed patients, such as cancer patients receiving hematopoietic cell transplants for the treatment of leukemia and other hematologic malignancies.[2] Aspergillosis is not limited to these patients though, as it occurs in other cases such as premature infants,[3] burn victims,[4] AIDS patients,[5] and patients in intensive care settings.[6] *Candida albicans*, as well as many other less prevalent fungal species, can infect the same patient populations.[2] *Coccidioides* spp. are responsible for valley fever (coccidioidomycosis), which is endemic in the southwestern United States, Mexico, and parts of South America, and can infect immunocompetent, but susceptible individuals.[7,8] Current antifungal drugs have limited effectiveness,[9] and *A. fumigatus*[10] and other species[11,12] are becoming resistant to our first-line treatments. Thus, the need exists for better control and prevention of fungal infections.

Experimental vaccines have shown considerable promise for the prevention of fungal infections. In *A. fumigatus* infection models, the protein Asp f3 (Pmp20) has been used to successfully prevent death

doi: 10.1111/j.1749-6632.2012.06761.x

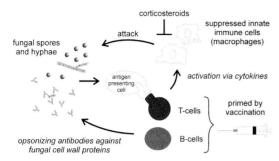

Figure 1. Proposed mechanism of antifungal T cell–based vaccines under conditions of corticosteroid immunosuppression.

of mice that were immunosuppressed with corticosteroids.[13,14] Similarly, a homolog of Asp f3, Pmp1 from *Coccidioides posadasii*, was effective in a murine model of coccidioidomycosis.[15] Another coccidioidal protein, Pep1, was also successful in protecting mice against *C. posadasii* challenge,[16] and Pep1 has an *A. fumigatus* homolog as well. Recently, the *A. fumigatus* protein Crf1 was shown to be cross-protective against experimental infections with *A. fumigatus* or *C. albicans*.[17] In addition, extracts from *Saccharomyces cerevisiae*, as well as whole heat-killed yeast cells, were shown in murine models to provide cross-protection against systemic *A. fumigatus* infection,[18,19] coccidioidomycosis,[20] candidiasis,[21] and cryptococcosis.[22]

Vaccine protection in these mouse models appears to function primarily via a Th1 mechanism, in which CD4$^+$ T cells are presented fungal antigens by antigen-presenting cells, such as dendritic cells or other phagocytes (Fig. 1). The activated T cells then stimulate phagocytes (primarily macrophages),[14,17,23] mainly via the cytokine interferon (IFN)-γ.[24] Activation with this cytokine increases the antifungal activity of the phagocytic effector cells and overcomes corticosteroid-induced immunosuppression.[25] Protective T cell transplants have been demonstrated successfully in immunosuppressed mice.[26] For example, protection was transferred from Asp f3–vaccinated animals into nonimmunized mice by transplantation of CD4$^+$ T cells, while transfer of Asp f3–specific antibodies was not protective.[14] However, antibodies could potentially enhance vaccine-derived immunity against fungi,[27,28] particularly when the targeted fungal antigen is located on the cell surface. Antibodies against such cell surface proteins could assist in opsonization of fungal cells by phagocytes, which in

turn would enhance the major histocompatibility complex presentation of specific antigens that can be recognized by vaccine-primed T cells (Fig. 1). In contrast, Asp f3 is an intracellular, peroxisomal protein not accessible to humoral antibodies, which may explain the lack of protection observed after transfer of Asp f3–specific antibodies into nonimmunized mice.[14]

An ideal vaccine would be cross protective against multiple species of fungi and would contain conserved homologous antigens (epitopes) in the fungal cell walls that elicit both antibody and T cell responses. Such antigens should also be as foreign to mammals as possible, to enhance their chance of being sufficiently immunogenic and to avoid triggering autoimmunity. To systematically discover promising potential vaccine candidates with these characteristics, we examined the total proteome of *A. fumigatus* using MS^E (mass spectrometry-elevated collision energy), a mass spectrometry technique allowing for label-free quantification of large numbers of proteins in complex samples.[29–33] Proteins identified from different fractions of *A. fumigatus* were quantified, and their subcellular protein localization was predicted. The amino acid sequences of promising candidates were compared to homologous proteins from other medically relevant fungal species, as well as mouse and human. *C. posadasii* hyphal protein extract was MS^E analyzed as well. Among filamentous fungi, *Coccidioides* is closely related to *Aspergillus*, and also in need of a vaccine.[28] Protection by the homologous vaccine candidates Asp f3[14] and Pmp1[15] against *A. fumigatus* and *C. posadasii*, respectively, indicates that *C. posadasii* may be a good target for finding potential vaccine candidates that are cross-protective against *A. fumigatus*.

Methods

Culture conditions

A. fumigatus strain AFCOH1 isolated from a patient at City of Hope (Duarte, California) and *C. posadasii* strain Silveira were used for all experiments. The AFCOH1 strain has continuously proven to be sufficiently virulent in animal models, and mass spectrometric analysis of its proteins has always matched available sequence data with excellent scoring quality. For growth of *A. fumigatus,* either Czapek Dox (CD) medium (a minimal medium, Difco, Detroit, MI) or potato dextrose (PD) medium (a rich medium, Difco, Detroit) was inoculated with 10^7 conidia/mL at 37 °C. Cultures were harvested in the late growth phase, i.e., after 24 h for PD medium and nine days for CD medium. *C. posadasii* was grown in a broth of 2% glucose and 1% yeast extract for five days at ambient temperature.

Sample preparation

A. fumigatus hyphal sonicate (HS) was prepared by cryogenic grinding of liquid nitrogen frozen dried hyphae with a MM301 ball mill (Retsch, Haan, Germany), followed by resuspension and sonication using a Sonicator 3000 (Misonix, Farmingdale, NY). *C. posadasii* hyphal extracts were prepared by collection of hyphae by filtration, suspension in phosphate buffered saline (PBS) and vortex mixing with glass beads. Homogenates were clarified by centrifugation and supernatants filter sterilized using 0.22 μm syringe filters. *A. fumigatus* culture filtrate (CF) was obtained by filtering spent fungal growth media through 0.22 μm PVDF pores, freezing the culture filtrate in liquid nitrogen, lyophilizing the frozen pellets, and filtering through Amicon centrifugal filter units with 10 kDa molecular weight cutoff (Millipore, Billerica, MA). The retentate was suspended to concentrate CF proteins 300×. Cell wall (CW) extract was obtained by digestion of hyphae for eight hours at 30 °C in a cell wall digestion cocktail containing 10 mg/mL 1-3-β-glucanase (Interspex, San Mateo, CA), 5 mg/mL Driselase (Interspex, San Mateo, CA), and 200 U/mL Lyticase (Sigma, St. Louis, MO), followed by centrifugation and filtration of the supernatant through a Millex PVDF filter with 0.22 μm pores (Millipore, Billerica, MA).

Mass spectrometry

Proteins were precipitated in 10% trichloroacetic acid, denatured with trifluoroethanol, reduced with tris-(2-carboxyethyl)-phosphine, alkylated with iodoacetamide, and digested overnight with trypsin (sequencing grade modified) (Promega, Madison, WI). Human serum albumin or rabbit glycogen phosphorylase was added as a standard to each sample for label-free quantification. MS^E was conducted using a SYNAPT G2 HD Q-TOF mass spectrometer, equipped with a UHPLC and an ion mobility 2D separation unit (Waters, Milford, MA). Data were analyzed and proteins tabulated using the Protein Lynx Global Server (PLGS) (Waters, Milford, MA).

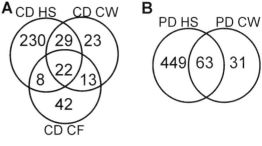

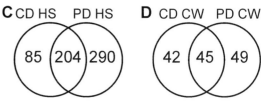

Figure 2. Comparison of proteins detected in different *A. fumigatus* fractions and growth media. Comparisons are among (A) HS, CF, and CW in CD medium, (B) HS and CW in PD medium, (C) HS from CD and PD media, and (D) CW from CD and PD media.

Protein analysis

When conducting cross-comparisons of different fungal fractions, proteins with a PLGS score of less than 500 were excluded from the analysis. Scaffold (Proteome Software, Portland, OR) was used for quantification of proteins. For homology analysis, NCBI's BLAST[34] was used to obtain percent identity and percent sequence alignment coverage between different proteins. Protein localization was predicted using the WoLF PSORT tool.[35]

Results

Mass spectrometry of *A. fumigatus* proteins

The 20 most abundant proteins from each *A. fumigatus* fraction were quantified using Scaffold, including HS/CD medium (Table S1), CF/CD medium (Table S2), CW/CD medium (Table S3), HS/PD medium (Table S4), and CW/PD medium (Table S5). Only proteins with WoLF PSORT-predicted extracellular localization are displayed in CW fractions.

In total, 367 proteins were detected in the three fractions of *A. fumigatus* grown in CD medium (Table S6). There were 289 proteins in the HS fraction, 85 in the CF fraction, and 87 in the CW fraction (Fig. 2A). Many proteins detected in the CW and HS fractions, but not CF, do not have predicted extracellular localization, and the bulk of these proteins

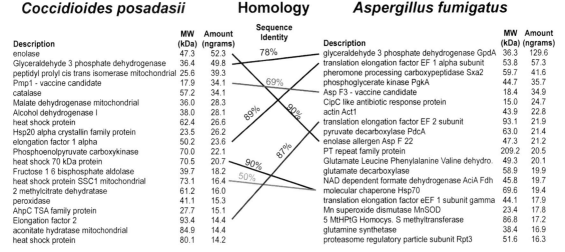

Figure 3. Homology between the 20 most abundant *A. fumigatus* and *C. posadasii* proteins. Quantification is per 1 μg of total protein.

in the CW fraction likely are contamination due to imperfect CW extraction methods. Of the most abundant 20 proteins in the CF, only one, XP˙749979.1, a protein of unknown function with cyanovirin-N homology (CVNH) domains, lacked extracellular localization according to WoLF PSORT. This small 11.8 kDa protein may lack a known extracellular signal. Furthermore, an independent study of the secreted proteome of *A. terreus* also identified a homolog of this protein.[36]

With *A. fumigatus* grown in PD medium, 494 proteins were detected in HS and 94 in CW fractions (Table S7); 63 proteins were detected in both fractions (Fig. 2B). Although HS from *A. fumigatus* grown in CD and PD contained many of the same proteins, a considerably greater number of proteins were detected in HS from growth in PD medium (Fig. 2C). This may be because the most abundant proteins in HS from growth in CD medium represented a greater proportion of the total protein than they did in HS from growth in PD medium. Although the majority of the 20 most abundant proteins in the CD and PD HS samples were detected in both samples, the 20 most abundant proteins in each HS sample had only four proteins in common. The number of proteins detected in CD and PD CW samples was approximately equal (Fig. 2D), with many of the proteins found in both samples. Of note, the abundant proteins major allergen Asp f2 (expressed during zinc-limiting conditions at higher pH[37]), catalase B, and L-amino acid oxidase LaoA

were not found in CW from growth in PD medium, and the abundant protein pheromone processing carboxypeptidase Sxa2 was not found in CW from growth in CD medium.

Proteomic comparison of *A. fumigatus* to *C. posadasii*

We detected 314 proteins via quantitative MS^E mass spectrometry in our *C. posadasii* extracts (Table S8), which was compared to the HS proteome *A. fumigatus*. The percent sequence identity of homologous regions of the 20 most abundant proteins in each fraction is shown in Figure 3. Five of the most abundant 20 *A. fumigatus* HS proteins had homologues in the top 20 *C. posadasii* HS proteins. Of note, the *A. fumigatus* protein Asp f3 (Pmp20), which has been successfully used to vaccinate mice against aspergillosis,[14] had a homolog in *C. posadasii* protein Pmp1 with 69% sequence identity. Pmp1 also protected mice against coccidioidomycosis when used as a vaccine.[15]

Interspecies protein homologies

Twenty cell wall proteins of interest were analyzed using NCBIs BLAST to determine potential homology with proteins from other medically relevant fungi, including *C. albicans, Cryptococcus neoformans, S. cerevisiae, Acremonium alcalophilum, Mucor circinelloides, Rhizopus oryzae, Penicillium marneffei, Fusarium oxysporum,* and *C. posadasii*, as well as homology with human and mouse

Table 1. Interspecies homology of select *A. fumigatus* cell wall proteins[a]

Protein	Accession[b]	kDa	Hs	Mm	Ca	Cn	Sc	Aa	Mc	Ro	Pm	Fo	Cp
Sxa2	XP_749486.1	59.8	29/56	36/27	28/87	60/100	27/81	29/76	28/85	25/84	64/100	40/93	47/90
All2	P79017.2	32.8	0	0	42/99	0	30/65	53/74	0	0	0	49/87	50/98
Crf1	Q8J0P4.2	40.3	0	33/26	48/66	32/56	49/62	48/67	38/43	34/42	67/65	59/67	49/66
Bys1	XP_748143.1	16.0	0	0	0	0	0	51/86	0	0	42/97	30/85	58/87
Ecm33	Q4WNS8.1	41.5	0	0	30/86	0	28/85	32/82	30/75	28/80	49/100	32/55	46/92
EglC	Q4WG16.1	44.7	0	0	27/61	0	30/56	57/65	26/54	25/62	65/70	62/65	28/61
GpiX[c]	XP_751929.1	39.6	0	0	0	0	0	0	0	0	55/100	0	49/45
Alp2	P87184.1	52.6	32/81	33/68	60/84	51/83	54/90	66/90	53/86	52/89	76/100	66/96	73/99
Gel4	P0C956.1	58.9	0	0	54/83	41/85	53/85	47/80	45/81	44/78	52/92	59/44	52/98
AbfB	Q4WL66.1	52.6	0	0	0	0	0	0	0	27/29	82/100	77/98	0
PHOa	Q8X176.1	49.1	0	0	0	0	0	0	43/72	45/66	66/87	52/78	0
CbhA	Q4WNA2.1	48.1	0	0	0	0	0	51/100	0	0	64/99	63/100	0
Al15	O60022.1	16.0	0	0	0	0	0	57/71	0	0	41/94	52/76	46/97
CpdS	XP_731524.1	57.1	29/70	28/70	29/69	41/85	30/68	26/95	31/76	28/66	69/91	45/85	52/85
CatB	Q92405.1	79.9	43/51	40/61	41/53	45/86	40/49	59/99	45/98	47/92	70/100	61/100	49/67
LaoA	XP_748830.1	78.6	30/26	25/72	0	0	0	35/98	35/96	0	0	72/80	25/46
Suc2p	XP_749260.1	57.3	0	0	49/97	49/63	0	0	0	0	0	48/97	0
Pep1	P41748.2	38.0	30/80	30/81	28/79	30/89	31/79	39/77	35/93	39/89	48/100	52/97	41/83
ChiA1	XP_747968.1	88.6	0	0	37/38	0	0	0	32/34	38/33	56/35	40/27	56/38
Crf2	Q4WI46.1	46.7	0	0	43/98	41/76	47/78	51/73	38/44	39/41	56/97	54/73	56/93

[a]Each cell lists % identity/% length of alignment against the most significant protein match. [b]XP accessions from NCBI REFSEQ. Others are from UniProtKB. [c]GpiX refers to an uncharacterized GPI-anchored cell wall protein. Hs, *Homo sapiens*; Mm, *Mus musculus*; Ca, *Candida albicans*; Cn, *Cryptococcus neofomans*; Sc, *Saccharomyces cerevisiae*; Aa, *Acremonium alcalophilum*; Mc, *Mucor circinelloides*; Ro, *Rhizopus oryzae*; Pm, *Penicillium marneffei*; Fo, *Fusarium oxysporum*; Cp, *Coccidiodes posadasii*

proteins (Table 1). Proteins with both % identity and percent coverage less than 25% were listed as 0, representing nonhomologous proteins. Many proteins were found to share high degrees of homology (>50% identity and >70% coverage) with proteins from other fungi, while retaining low homology (<35% identity and <30% coverage) with human or mouse proteins. Such proteins are potential broad-spectrum antifungal vaccine candidates. Crf1, a highly abundant *A. fumigatus* CW protein, has already shown cross-protection against *A. fumigatus* and *C. albicans* in mouse models of infection, demonstrating the feasibility of a cross-protective vaccine.[17]

Discussion

Although several non-cell wall proteins existed in our *A. fumigatus* CW fractions (of which NAD-dependent malate dehydrogenase, ubiquitin, spermidine synthase, cytochrome b5, and a hypothetical protein were the most abundant), cell wall proteins did appear to be highly enriched in these fractions compared to other proteins. Over one-half of the most abundant proteins in this fraction have predicted extracellular localization, which includes cell wall proteins. Several proteins with predicted extracellular localization were also detected in CF, but such proteins were usually also detected in HS as well, and many of these proteins have known cell wall localization. Furthermore, many abundant CF and HS proteins were not detected in CW fractions. This implies that any noncell wall protein contaminants in our CW fractions are only a particular subfraction of other *A. fumigatus* proteins. Thus, we may hypothesize that certain abundant cell wall proteins may ultimately be released into the extracellular environment, either to serve a specific function or to simply be lost during cell wall remodeling. Finally, several proteins lacking extracellular localization prediction from WoLF PSORT may in fact still be cell wall proteins with unknown export and translocation pathways. A study of the conidial cell wall proteome of *A. fumigatus* revealed similar proteins lacking classical signal peptides.[38] The proteins discovered in conidia have very little overlap with the cell wall proteins we encountered in hyphae, implying that the cell wall proteome undergoes major changes between the conidia and hyphae stages of the fungal lifecycle. This change in the cell wall proteome may result in an antifungal vaccine against one morphologic form being ineffective against another form (i.e., a vaccine against conidial cell wall might not induce protection against the hyphal cell wall). It may therefore be desirable to develop vaccines using antigens expressed in both early and later stages of the fungal lifecycle. Our protein identifications from HS fractions were very similar to previous studies.[39,40] One study on the secreted proteome of *A. fumigatus* discovered many proteins detected in our CF fraction, but also several proteins with intracellular localization.[41]

We found the cell wall proteome of *A. fumigatus* to be rich in proteins related to carbohydrate digestion and cell wall structure formation and maintenance. Indeed, the 20 most abundant proteins from growth in CD and PD media CW fractions contained many proteins in this category, including Crf1, Bys1, EglC, Ecm33, CbhA, AbfB, Gel4, and Gel1. Several other less abundant proteins of this category were detected in the CW fraction from growth in each medium type individually, as were several proteins related to protein digestion and processing. Although the cell wall proteome was very similar during growth in CD or PD medium, there were significant differences in expression of proteins related to peptide processing and digestion. Asp f2 (All2) and LaoA were found only in CW from growth in CD medium, while Sxa2 was found only from growth in PD CW fraction. PD contains tryptic digest of casein, and CD contains no peptides, which could account for this difference.

When determining optimal proteins for vaccine candidates against fungal infection, it is essential to ensure that such a protein is expressed at high levels during pulmonary infection, because lung infections are most relevant and severe. Thus, proteins that are expressed under growth conditions simulating both a rich and a minimal supply of nutrients should be considered the best candidates for vaccine trials. Ideally, the *A. fumigatus* proteome could be assessed under conditions similar to those found during lung infection.

A high level of homology between proteins in different fungal species is another essential aspect of a cross-protective vaccine candidate. Our study comparing *A. fumigatus* and *C. posadasii* hyphal extracts indicates that several abundantly expressed proteins have high degrees of homology between these species. The same argument may apply to fungal cell wall proteins, and our preliminary data on

the cell wall proteome of *A. terreus* agrees with this notion (data not shown). If these proteins have identical or highly similar and protective epitopes, they could serve as cross-protective vaccines. The most promising candidates would have relatively low homology to human and mouse proteins, and also high homology to proteins from *C. albicans*, second only to *A. fumigatus* as a cause of lethal invasive fungal infections. Therefore, our best potential cross-protective vaccine candidates include Crf1, Ecm33, Alp2, Gel4, and Crf2.

Animal vaccination experiments are currently planned or underway using recombinant Sxa2, Asp f2, Crf1, Bys1, Ecm33, EglC, GpiX, Alp2, Gel4, PHOa, and Crf2. Also planned is a broader mass spectrometry study to compare the entire proteome of many different species of pathogenic fungi, including filamentous fungi and yeast.

Acknowledgments

This work was supported, in part, by the Hermann Foundation and the Tim Nesvig Lymphoma Fellowship and Research Fund.

Conflicts of interest

The authors declare no conflicts of interest.

Supporting Information

Additional Supporting Information may be found in the online version of this article.

Table S1. Most abundant 20 proteins in hyphae sonicate of *A. fumigatus* grown in Czapek Dox medium.

Table S2. Most abundant 20 proteins in culture filtrate of *A. fumigatus* grown in Czapek Dox medium.

Table S3. Most abundant 20 proteins in cell wall of *A. fumigatus* grown in Czapek Dox medium.

Table S4. Most abundant 20 proteins in hyphae sonicate of *A. fumigatus* grown in Potato Dextrose medium.

Table S5. Most abundant 20 proteins in cell wall of *A. fumigatus* grown in Potato Dextrose medium.

Table S6. List of proteins in hyphae sonicate (HS), culture filtrate (CF), and cell wall (CW) extracts of *A. fumigatus* grown in Czapek Dox medium. PLGS (Protein Lynx Global Server) score is a measure of the quality of the mass spectrometry–based protein identification.

Table S7. List of proteins in hyphae sonicate (HS) and cell wall (CW) extracts of *A. fumigatus* grown in Potato Dextrose medium. PLGS (Protein Lynx Global Server) score is a measure of the quality of a mass spectrometry–based protein identification.

Table S8. List of proteins in hyphae extract of *C. posadasii.*

References

1. Singh, N. & D.L. Paterson. 2005. *Aspergillus* infections in transplant recipients. *Clin. Microbiol. Rev.* **18:** 44–69.
2. Caston-Osorio, J.J., A. Rivero & J. Torre-Cisneros. 2008. Epidemiology of invasive fungal infection. *Int. J. Antimicrob. Agents* **32(Suppl 2):** S103–S109.
3. Dotis, J., E. Iosifidis & E. Roilides. 2007. Central nervous system aspergillosis in children: a systematic review of reported cases. *Int. J. Infect. Dis.* **11:** 381–393.
4. Chakrabarti, A., V. Gupta, G. Biswas, *et al.* 1998. Primary cutaneous aspergillosis: our experience in 10 years. *J. Infect.* **37:** 24–27.
5. Warnock, D.W. 2007. Trends in the epidemiology of invasive fungal infections. *Nippon. Ishinkin. Gakkai. Zasshi.* **48:** 1–12.
6. Vandewoude, K., D. Vogelaers & S. Blot. 2006. Aspergillosis in the ICU—The new 21st century problem? *Med. Mycol.* **44:** S71–S76.
7. Cox, R.A. & D.M. Magee. 2004. Coccidioidomycosis: host response and vaccine development. *Clin. Microbiol. Rev.* **17:** 804–839, table of contents.
8. Hector, R.F. & R. Laniado-Laborin. 2005. Coccidioidomycosis—a fungal disease of the Americas. *PLoS Med.* **2:** e2.
9. Dagenais, T.R. & N.P. Keller. 2009. Pathogenesis of *Aspergillus fumigatus* in invasive aspergillosis. *Clin. Microbiol. Rev.* **22:** 447–465.
10. Verweij, P.E., S.J. Howard, W.J. Melchers & D.W. Denning. 2009. Azole-resistance in *Aspergillus*: proposed nomenclature and breakpoints. *Drug Resist. Updat.* **12:** 141–147.
11. Malani, A.N. & C.A. Kauffman. 2007. Changing epidemiology of rare mold infections: implications for therapy. *Drugs* **67:** 1803–1812.
12. Chen, S.C., E.G. Playford & T.C. Sorrell. 2010. Antifungal therapy in invasive fungal infections. *Curr. Opin. Pharmacol.* **10:** 522–530.
13. Ito, J.I., J.M. Lyons, T.B. Hong, *et al.* 2006. Vaccinations with recombinant variants of *Aspergillus fumigatus* allergen Asp f 3 protect mice against invasive aspergillosis. *Infect. Immun.* **74:** 5075–5084.
14. Diaz-Arevalo, D., K. Bagramyan, T.B. Hong, *et al.* 2011. CD4+ T cells mediate the protective effect of the recombinant Asp f3-based anti-aspergillosis vaccine. *Infect. Immun.* **79:** 2257–2266.
15. Orsborn, K.I., L.F. Shubitz, T. Peng, *et al.* 2006. Protein expression profiling of *Coccidioides posadasii* by two-dimensional differential in-gel electrophoresis and evaluation of a newly recognized peroxisomal matrix protein as a recombinant vaccine candidate. *Infect. Immun.* **74:** 1865–1872.
16. Tarcha, E.J., V. Basrur, C.Y. Hung, *et al.* 2006. A recombinant aspartyl protease of *Coccidioides posadasii* induces protection against pulmonary coccidioidomycosis in mice. *Infect. Immun.* **74:** 516–527.
17. Stuehler, C., N. Khanna, S. Bozza, *et al.* 2011. Cross-protective TH1 immunity against *Aspergillus fumigatus* and *Candida albicans. Blood* **117:** 5881–5891.
18. Liu, M., J. Capilla, M.E. Johansen, *et al.* 2011. *Saccharomyces* as a vaccine against systemic aspergillosis: 'the friend of man' a friend again? *J. Med. Microbiol.* **60:** 1423–1432.
19. Stevens, D.A., K.V. Clemons & M. Liu. 2011. Developing a vaccine against aspergillosis. *Med. Mycol.* **49(Suppl. 1):** S170–S176.
20. Capilla, J., K.V. Clemons, M. Liu, *et al.* 2009. *Saccharomyces cerevisiae* as a vaccine against coccidioidomycosis. *Vaccine* **27:** 3662–3668.
21. Liu, M., K.V. Clemons, M.E. Johansen, *et al.* 2012. *Saccharomyces* as a vaccine against systemic candidiasis. *Immunol. Invest.*, DOI 10.3109/08820139.2012.692418.
22. Majumder, T., M. Liu, V. Chen, *et al.* 2011. Killed *Saccharomyces cerevisiae* protects against lethal challenge with *Cryptococcus grubii. 51st Interscience Conference on Antimicrobial Agents and Chemotherapy*, Chicago Abstracts, no. G1–769.
23. Fierer, J., C. Waters & L. Walls. 2006. Both CD4+ and CD8+ T cells can mediate vaccine-induced protection against *Coccidioides immitis* infection in mice. *J. Infect. Dis.* **193:** 1323–1331.
24. Hebart, H., C. Bollinger, P. Fisch, *et al.* 2002. Analysis of T-cell responses to *Aspergillus fumigatus* antigens in healthy individuals and patients with hematologic malignancies. *Blood* **100:** 4521–4528.
25. Beck, O., M.S. Topp, U. Koehl, *et al.* 2006. Generation of highly purified and functionally active human TH1 cells against *Aspergillus fumigatus. Blood* **107:** 2562–2569.
26. Lehrnbecher, T., O. Beck, U. Koehl & L. Tramsen. 2009. Cultivated anti-*Aspergillus* T(H)1 cells. *Med. Mycol.* **47(Suppl 1):** S170–S174.
27. Montagnoli, C., S. Bozza, A. Bacci, *et al.* 2003. A role for antibodies in the generation of memory antifungal immunity. *Eur. J. Immunol.* **33:** 1193–1204.
28. Clemons, K.V. & D.A. Stevens. 2011. Vaccines against coccidioidomycosis: past and current developments. *J. Invasive Fungal Infect.* **5:** 99–109.
29. Cheng, F.Y., K. Blackburn, Y.M. Lin, *et al.* 2009. Absolute protein quantification by LC/MS(E) for global analysis of salicylic acid-induced plant protein secretion responses. *J. Proteome Res.* **8:** 82–93.
30. Bostanci, N., W. Heywood, K. Mills, *et al.* 2010. Application of label-free absolute quantitative proteomics in human gingival crevicular fluid by LC/MS E (gingival exudatome). *J. Proteome Res.* **9:** 2191–2199.
31. Li, G.Z., J.P. Vissers, J.C. Silva, *et al.* 2009. Database searching and accounting of multiplexed precursor and product ion spectra from the data independent analysis of simple and complex peptide mixtures. *Proteomics* **9:** 1696–1719.
32. Geromanos, S.J., J.P. Vissers, J.C. Silva, *et al.* 2009. The detection, correlation, and comparison of peptide precursor and product ions from data independent LC-MS with data dependant LC-MS/MS. *Proteomics* **9:** 1683–1695.

33. Vissers, J.P., S. Pons, A. Hulin, *et al.* 2009. The use of proteome similarity for the qualitative and quantitative profiling of reperfused myocardium. *J. Chromatogr B Analyt. Technol. Biomed. Life Sci.* **877:** 1317–1326.

34. Altschul, S.F., W. Gish, W. Miller, *et al.* 1990. Basic local alignment search tool. *J. Mol. Biol.* **215:** 403–410.

35. Horton, P., K.J. Park, T. Obayashi, *et al.* 2007. WoLF PSORT: protein localization predictor. *Nucleic Acids Res.* **35:** W585–W587.

36. Han, M.J., N.J. Kim, S.Y. Lee & H.N. Chang. 2010. Extracellular proteome of *Aspergillus terreus* grown on different carbon sources. *Curr. Genet.* **56:** 369–382.

37. Amich, J., R. Vicentefranqueira, F. Leal & J.A. Calera. 2010. *Aspergillus fumigatus* survival in alkaline and extreme zinc-limiting environments relies on the induction of a zinc homeostasis system encoded by the zrfC and aspf2 genes. *Eukaryot. Cell* **9:** 424–437.

38. Asif, A.R., M. Oellerich, V.W. Amstrong, *et al.* 2006. Proteome of conidial surface associated proteins of *Aspergillus fumigatus* reflecting potential vaccine candidates and allergens. *J. Proteome Res.* **5:** 954–962.

39. Teutschbein, J., D. Albrecht, M. Potsch, *et al.* 2010. Proteome profiling and functional classification of intracellular proteins from conidia of the human-pathogenic mold *Aspergillus fumigatus. J. Proteome Res.* **9:** 3427–3442.

40. Vodisch, M., D. Albrecht, F. Lessing, *et al.* 2009. Two-dimensional proteome reference maps for the human pathogenic filamentous fungus *Aspergillus fumigatus. Proteomics* **9:** 1407–1415.

41. Singh, B., M. Oellerich, R. Kumar, *et al.* 2010. Immunoreactive molecules identified from the secreted proteome of *Aspergillus fumigatus. J. Proteome Res.* **9:** 5517–5529.

Ann. N.Y. Acad. Sci. ISSN 0077-8923

ANNALS OF THE NEW YORK ACADEMY OF SCIENCES
Issue: *Advances Against Aspergillosis*

Inflammation in aspergillosis: the good, the bad, and the therapeutic

Agostinho Carvalho, Cristina Cunha, Rossana G. Iannitti, Antonella De Luca, Gloria Giovannini, Francesco Bistoni, and Luigina Romani

Department of Experimental Medicine and Biochemical Sciences, University of Perugia, Perugia, Italy

Address for correspondence: Luigina Romani, Department of Experimental Medicine and Biochemical Sciences, University of Perugia, Via del Giochetto, 06122 Perugia, Italy. lromani@unipg.it

Aspergillosis includes a spectrum of diseases caused by different *Aspergillus* spp. New insights into the cellular and molecular mechanisms of resistance and immune tolerance to the fungus in infection and allergy have been obtained in experimental settings. The fact that virulence factors, traditionally viewed as fungal attributes, are contingent upon microbial adaptation to various environmental stresses encountered in the human host implies that the host and fungus are jointly responsible for pathogenicity. Ultimately, despite the occurrence of severe aspergillosis in immunocompromised patients, clinical evidence indicates that aspergillosis also occurs in the setting of a heightened inflammatory response, in which immunity occurs at the expense of host damage and pathogen eradication. Thus, targeting pathogenicity rather than microbial growth, tolerance rather than resistance mechanisms of defense may pave the way to targeted anti-inflammatory strategies in difficult-to-treat patients. The challenge now is to translate promising results from experimental models to the clinic.

Keywords: *Aspergillus*; inflammation; Th cells; PAMPs; DAMPs; tolerance

Introduction

Aspergillus is a large genus containing more than 180 species that are ubiquitous in nature. Among the species, *Aspergillus fumigatus* remains the most common cause of human diseases, followed by other species including A. flavus, A. niger, A. nidulans, and A. terreus. Diseases caused by *Aspergillus* include saprophytic colonization of preexisting cavities (aspergilloma), allergic asthma, hypersensitivity pneumonitis, allergic bronchopulmonary aspergillosis (ABPA) occurring as a complication of bronchial asthma or cystic fibrosis, and invasive aspergillosis. Invasive aspergillosis is associated with a high mortality rate in patients with dysregulated immunity, such as patients with hematological malignancies, recipients of solid organs and stem cell transplants, and patients admitted to the intensive care unit.[1] Although the past decade has witnessed significant progress in the management of invasive aspergillosis, the infection continues to be a deadly disease.[1]

The main reasons for this include intrinsic or acquired antifungal resistance and the deleterious effect of a dysregulated inflammation.

Current understanding of the pathophysiology underlying *Aspergillus* infection and disease highlights the multiple cell populations and cell-signaling pathways involved in these complex conditions. Applying systems biology approaches to these complex processes has resulted in a better appreciation of the intricate cross talk provided by temporal changes in mediators, metabolites, and cell phenotypes underlining the coordinated processes.[2] In addition to the pathogen-derived ligands (PAMPs) of pattern recognition receptors (PRRs),[3] primary metabolites associated with fungal growth and adaptation[4] and damage-associated molecular patterns (DAMPs)[5] all contribute to fungal pathogenesis and generation of antifungal immune response. These emerging themes in fungal pathogenesis better accommodate the susceptibility to fungal diseases in primary immunodeficiencies,

doi: 10.1111/j.1749-6632.2012.06754.x

 Ann. N.Y. Acad. Sci. 1273 (2012) 52–59 © 2012 New York Academy of Sciences.

Restraining pathogenic inflammation by exogenous kynurenines

In experimental CGD, IL-17A neutralization increased fungal clearance, ameliorated inflammatory pathology, and restored protective Th1 antifungal resistance.[7] Perhaps more importantly, complete cure and reversal of the hyperinflammatory CGD phenotype were achieved by administration of supplemental L-kynurenine, an early amino acid catabolite of L-tryptophan in the IDO-dependent pathway.[7] Similar results have been obtained in experimental CF.[8]

Targeting IDO in fungal allergy

Consistent with the responsiveness of ABPA to corticosteroid treatment,[52] the Th2 allergic phenotype is attenuated by dexamethasone, which enhanced production of IL-10 and Foxp3 transcripts, both markers of protective T_{reg} cell activity in *Aspergillus* allergy. These data demonstrated that dexamethasone downregulates exacerbating Th2 cell responses in ABPA by inhibiting the expansion and activation of Th2 cells and by upregulating the expression of Foxp3 via mechanisms that require tryptophan catabolism.[50] Thus, induction of IDO could be an important mechanism underlying the anti-inflammatory action of corticosteroids.

Targeting inflammatory pathways via siRNA

It has been demonstrated that the activation of distinct signaling pathways in DCs and other antigen-presenting cells translates recognition of the fungus into distinct inflammatory and adaptive immune responses[42] (Fig. 1). Thus, the screening of signaling pathways in DCs through a systems biology approach could be exploited for the development of therapeutics to attenuate inflammation in respiratory fungal infections and diseases. Indeed, the study showed the utility of the DC systems biology approach for quick screening of therapeutic siRNA to attenuate inflammation in *Aspergillus* infection and allergy. The intranasal administration of siRNA has opened new avenues in drug delivery and respiratory therapy.[53]

Exploiting IDO + DCs as fungal vaccines in transplantation

The finding that *Aspergillus*-pulsed DCs could be used as a vaccine in hematopoietic transplantation[54] raised excitement about vaccination against aspergillosis.[55] Within the instructive model of DC-mediated regulation of the Th repertoire, it is conceivable that an improved understanding of the pathogen/DC interaction will allow the potential use of pathogen- or TLR-conditioned DCs for the induction of patient-tailored T_{reg} cells with indirect antidonor allospecificity. Over recent years, experimental models have shown that it is possible to exploit the mechanisms that normally maintain immune homeostasis and tolerance to self-antigens to induce tolerance to alloantigens.[56] Like natural tolerance, transplantation tolerance is achieved through control of T cell reactivity by central and peripheral mechanisms of tolerance. We have recently found that this goal is achievable by the adoptive cellular therapy of *Aspergillus*-pulsed DCs + IDO that could induce antifungal resistance within a regulatory environment.[57] Tolerogenic DCs + IDO proved to be pivotal in the generation of some form of dominant regulation that ultimately controlled inflammation, pathogen immunity, and tolerance in transplant recipients, eventually leading to prevention of graft-versus-host reaction and reduction of aspergillosis incidence rates.

Conclusions

In summary, because immunity is neither a set of discrete signaling pathways activated by pathogens nor simply a function of the host, the use of systems biology approaches has proved to be useful for the generation of new therapeutics that targets pathogenicity rather than microbial growth, targets the host–pathogen interface rather than the pathogen, and promotes protective immune responses. It is now clear that genetic variants of molecules involved in innate recognition of fungi may account, in part, for the inherited differences in human susceptibility to fungal infections.[58,59] Thus, the integration of systems biology approaches with immunogenetic screenings may promote the development of targeted immunotherapeutic strategies in high-risk patients.

Acknowledgments

We thank Dr. Cristina Massi Benedetti for digital art and editing. This work was supported by the Specific Targeted Research Project ALLFUN (FP7–HEALTH–2009 Contract Number 260338) and the Italian Grant Application 2010 Fondazione per la Ricerca sulla Fibrosi Cistica (Research Project FFC#21/2010), with the contribution of the

Francesca Guadagnin, Coca Cola Light Tribute to Fashion and Delegazione FFC di Belluno.

Conflicts of interest

The authors declare no conflicts of interest.

References

1. Latge, J.P. & W.J. Steinbach. 2009. *Aspergillus fumigatus and Aspergillosis.* ASM Press. Washington, DC.
2. Santamaria, R. *et al.* 2011. Systems biology of infectious diseases: a focus on fungal infections. *Immunobiology* **216:** 1212–1227.
3. Romani, L. 2011. Immunity to fungal infections. *Nat. Rev. Immunol.* **11:** 275–288.
4. Grahl, N. *et al.* 2011. *In vivo* hypoxia and a fungal alcohol dehydrogenase influence the pathogenesis of invasive pulmonary aspergillosis. *PLoS Pathog.* **7:** e1002145.
5. Sorci, G. *et al.* 2011. The danger signal S100B integrates pathogen- and danger-sensing pathways to restrain inflammation. *PLoS Pathog.* **7:** e1001315.
6. De Luca, A. *et al.* 2012. CD4 T cell vaccination overcomes defective cross-presentation of fungal antigens in murine chronic granulomatous disease. *J. Clin. Invest.* **122:** 1816–1831.
7. Romani, L. *et al.* 2008. Defective tryptophan catabolism underlies inflammation in mouse chronic granulomatous disease. *Nature* **451:** 211–215.
8. Iannitti, R.G. *et al.* 2012. Aspergillosis in cystic fibrosis: a multifactorial disease? In *Proceedings of the 5th Advances Against Aspergillosis Conference.* Istanbul, Turkey.
9. Perfect, J.R. 2012. The impact of the host on fungal infections. *Am. J. Med.* **125:** S39–S51.
10. Romani, L. & P. Puccetti. 2006. Protective tolerance to fungi: the role of IL-10 and tryptophan catabolism. *Trends Microbiol.* **14:** 183–189.
11. Gupta, A.O. & N. Singh. 2011. Immune reconstitution syndrome and fungal infections. *Curr. Opin. Infect. Dis.* **24:** 527–533.
12. Antachopoulos, C., T.J. Walsh & E. Roilides. 2007. Fungal infections in primary immunodeficiencies. *Eur. J. Pediatr.* **166:** 1099–1117.
13. Holland, S.M. *et al.* 2007. STAT3 mutations in the hyper-IgE syndrome. *N. Engl. J. Med.* **357:** 1608–1619.
14. Ortega, M. *et al.* 2006. Prospective evaluation of procalcitonin in adults with non-neutropenic fever after allogeneic hematopoietic stem cell transplantation. *Bone Marrow Transplant* **37:** 499–502.
15. Schubert, M.S. 2006. Allergic fungal sinusitis. *Clin. Rev. Allergy Immunol.* **30:** 205–216.
16. Park, S.J. & B. Mehrad. 2009. Innate immunity to *Aspergillus* species. *Clin. Microbiol. Rev.* **22:** 535–551.
17. Balloy, V. & M. Chignard. 2009. The innate immune response to *Aspergillus fumigatus. Microbes Infect.* **11:** 919–927.
18. Li, Z.Z. *et al.* 2012. Role of NOD2 in regulating the immune response to *Aspergillus fumigatus. Inflamm. Res.* **61:** 643–648.
19. Said-Sadier, N. *et al.* 2010. *Aspergillus fumigatus* stimulates the NLRP3 inflammasome through a pathway requiring ROS production and the Syk tyrosine kinase. *PLoS One* **5:** e10008.
20. Latge, J.P. 2010. Tasting the fungal cell wall. *Cell. Microbiol.* **12:** 863–872.
21. Balloy, V. *et al.* 2005. Differences in patterns of infection and inflammation for corticosteroid treatment and chemotherapy in experimental invasive pulmonary aspergillosis. *Infect. Immun.* **73:** 494–503.
22. Romani, L. & P. Puccetti. 2007. Controlling pathogenic inflammation to fungi. *Expert Rev. Anti Infect. Ther.* **5:** 1007–1017.
23. Singh, N. & J.R. Perfect. 2007. Immune reconstitution syndrome associated with opportunistic mycoses. *Lancet Infect. Dis.* **7:** 395–401.
24. Bignell, E. *et al.* 2005. Virulence comparisons of *Aspergillus nidulans* mutants are confounded by the inflammatory response of p47phox-/- mice. *Infect. Immun.* **73:** 5204–5207.
25. Moretti, S. *et al.* 2008. The contribution of PARs to inflammation and immunity to fungi. *Mucosal Immunol.* **1:** 156–168.
26. Cunha, C. *et al.* 2011. Genetically-determined hyperfunction of the S100B/RAGE axis is a risk factor for aspergillosis in stem cell transplant recipients. *PLoS One* **6:** e27962.
27. Cooney, N.M. & B.S. Klein. 2008. Fungal adaptation to the mammalian host: it is a new world, after all. *Curr. Opin. Microbiol.* **11:** 511–516.
28. Barker, B.M. *et al.* 2012. Transcriptomic and proteomic analyses of the *Aspergillus fumigatus* hypoxia response using an oxygen-controlled fermenter. *BMC Genomics.* **13:** 62.
29. Grahl, N. *et al.* 2012. Hypoxia and fungal pathogenesis: to air or not to air? *Eukaryot Cell* **11:** 560–570.
30. Murdock, B.J. *et al.* 2011. Coevolution of TH1, TH2, and TH17 responses during repeated pulmonary exposure to *Aspergillus fumigatus* conidia. *Infect. Immun.* **79:** 125–135.
31. Chaudhary, N., J.F. Staab & K.A. Marr. 2010. Healthy human T-cell responses to *Aspergillus fumigatus* antigens. *PLoS One* **5:** e9036.
32. Potenza, L. *et al.* 2008. Assessment of *Aspergillus*-specific T cells for diagnosis of invasive aspergillosis in a leukemic child with liver lesions mimicking hepatosplenic candidiasis. *Clin. Vaccine Immunol.* **15:** 1625–1628.
33. Cohen, N.R. *et al.* 2011. Innate recognition of cell wall beta-glucans drives invariant natural killer T cell responses against fungi. *Cell Host Microbe.* **10:** 437–450.
34. Filipe-Santos, O. *et al.* 2006. Inborn errors of IL-12/23- and IFN-gamma-mediated immunity: molecular, cellular, and clinical features. *Semin. Immunol.* **18:** 347–361.
35. Porter, P.C. *et al.* 2011. Necessary and sufficient role for T helper cells to prevent fungal dissemination in allergic lung disease. *Infect. Immun.* **79:** 4459–4471.
36. Allard, J.B. *et al.* 2006. *Aspergillus fumigatus* generates an enhanced Th2-biased immune response in mice with defective cystic fibrosis transmembrane conductance regulator. *J. Immunol.* **177:** 5186–5194.
37. Mueller, C. *et al.* 2011. Lack of cystic fibrosis transmembrane conductance regulator in CD3+ lymphocytes leads to aberrant cytokine secretion and hyperinflammatory adaptive immune responses. *Am. J. Respir. Cell Mol. Biol.* **44:** 922–929.
38. Kreindler, J.L. *et al.* 2010. Vitamin D3 attenuates Th2 responses to *Aspergillus fumigatus* mounted by CD4+ T cells

from cystic fibrosis patients with allergic bronchopulmonary aspergillosis. *J. Clin. Invest.* **120:** 3242–3254.

39. Murdock, B.J. *et al.* 2012. Interleukin-17 drives pulmonary eosinophilia following repeated exposure to *Aspergillus fumigatus Conidia. Infect. Immun.* **80:** 1424–1436.

40. Moreira, A.P. *et al.* 2011. The protective role of TLR6 in a mouse model of asthma is mediated by IL-23 and IL-17A. *J. Clin. Invest.* **121:** 4420–4432.

41. Zelante, T. *et al.* 2007. IL-23 and the Th17 pathway promote inflammation and impair antifungal immune resistance. *Eur. J. Immunol.* **37:** 2695–2706.

42. Bonifazi, P. *et al.* 2010. Intranasally delivered siRNA targeting PI3K/Akt/mTOR inflammatory pathways protects from aspergillosis. *Mucosal Immunol.* **3:** 193–205.

43. Chai, L.Y. *et al.* 2010. Anti-*Aspergillus* human host defence relies on type 1 T helper (Th1), rather than type 17 T helper (Th17), cellular immunity. *Immunology* **130:** 46–54.

44. Montagnoli, C. *et al.* 2006. Immunity and tolerance to *Aspergillus* involve functionally distinct regulatory T cells and tryptophan catabolism. *J. Immunol.* **176:** 1712–1723.

45. Roilides, E. *et al.* 1998. Increased serum concentrations of interleukin-10 in patients with hepatosplenic candidiasis. *J. Infect. Dis.* **178:** 589–592.

46. Hebart, H. *et al.* 2002. Analysis of T-cell responses to *Aspergillus fumigatus* antigens in healthy individuals and patients with hematologic malignancies. *Blood* **100:** 4521–4528.

47. Sambatakou, H. *et al.* 2006. Cytokine profiling of pulmonary aspergillosis. *Int. J. Immunogenet.* **33:** 297–302.

48. Puccetti, P. & U. Grohmann. 2007. IDO and regulatory T cells: a role for reverse signalling and non-canonical NF-kappaB activation. *Nat. Rev. Immunol.* **7:** 817–823.

49. von Bubnoff, D. *et al.* 2004. Asymptomatic atopy is associated with increased indoleamine 2,3-dioxygenase activity and interleukin-10 production during seasonal allergen exposure. *Clin. Exp. Allergy* **34:** 1056–1063.

50. Grohmann, U. *et al.* 2007. Reverse signaling through GITR ligand enables dexamethasone to activate IDO in allergy. *Nat. Med.* **13:** 579–586.

51. De Luca, A. *et al.* 2010. Non-hematopoietic cells contribute to protective tolerance to *Aspergillus fumigatus* via a TRIF pathway converging on IDO. *Cell. Mol. Immunol.* **7:** 459–470.

52. Judson, M.A. & D.A. Stevens. 2001. Current pharmacotherapy of allergic bronchopulmonary aspergillosis. *Expert Opin. Pharmacother.* **2:** 1065–1071.

53. Bitko, V. & S. Barik. 2008. Nasal delivery of siRNA. *Methods Mol. Biol.* **442:** 75–82.

54. Bozza, S. *et al.* 2003. A dendritic cell vaccine against invasive aspergillosis in allogeneic hematopoietic transplantation. *Blood* **102:** 3807–3814.

55. Stevens, D.A. 2004. Vaccinate against aspergillosis! A call to arms of the immune system. *Clin. Infect. Dis.* **38:** 1131–1136.

56. Waldmann, H. & S. Cobbold. 2004. Exploiting tolerance processes in transplantation. *Science* **305:** 209–212.

57. Montagnoli, C. *et al.* 2008. Provision of antifungal immunity and concomitant alloantigen tolerization by conditioned dendritic cells in experimental hematopoietic transplantation. *Blood Cells Mol. Dis.* **40:** 55–62.

58. Carvalho, A. *et al.* 2009. Polymorphisms in Toll-like receptor genes and susceptibility to infections in allogeneic stem cell transplantation. *Exp. Hematol.* **37:** 1022–1029.

59. Mezger, M., H. Einsele & J. Loeffler. 2010. Genetic susceptibility to infections with *Aspergillus fumigatus. Crit. Rev. Microbiol.* **36:** 168–177.

Ann. N.Y. Acad. Sci. ISSN 0077-8923

ANNALS OF THE NEW YORK ACADEMY OF SCIENCES
Issue: *Advances Against Aspergillosis*

Pattern recognition receptors and their role in invasive aspergillosis

Mark S. Gresnigt, Mihai G. Netea, and Frank L. van de Veerdonk

Department of Medicine, Radboud University, Nijmegen Medical Center, and Nijmegen Institute for Infection, Inflammation, and Immunity (N4i), Nijmegen, the Netherlands

Address for correspondence: Frank van de Veerdonk, Geert Grooteplein Zuid 8, 6525 GA Nijmegen, the Netherlands. f.veerdonk@aig.umcn.nl

Pattern recognition receptors (PRRs) are germline receptors that recognize conserved structures on microorganisms. Several PRRs have been identified in the recent years that are involved in the immune response against *Aspergillus fumigatus*. The role of PRRs in invasive pulmonary aspergillosis becomes especially apparent in the setting of an immunocompromised status of the host because of the redundancy of many PRRs in the host defense against *A. fumigatus*. Studies that investigated the PRRs and their effector pathways in invasive aspergillosis have led to a better understanding of the pathogenesis of invasive aspergillosis in immunocompromised patients. This knowledge may pave the way for novel diagnostic and immunomodulatory treatment strategies that are needed to overcome the high mortality associated with invasive *A. fumigatus* infection in immunocompromised patients.

Keywords: *Aspergillus fumigatus*; Toll-like receptors; C-type lectin receptors; NOD-like receptors; innate immune cells

Introduction

Aspergillus fumigatus can cause invasive pulmonary aspergillosis in immunocompromised patients. The management of invasive aspergillosis still presents a major challenge, and knowledge of the host defense against *A. fumigatus* is essential to develop new treatment strategies. Here, we review the literature on pattern recognition receptors (PRRs) that recognize *A. fumigatus*, and focus on their role in the host defense against invasive pulmonary aspergillosis (IPA).

Pattern recognition of *A. fumigatus*

The first step in the innate host defense against *A. fumigatus* is the recognition of fungal cell wall components by PRRs present on and in the cells of the innate immune system. Three of the four major families of PRR have been implicated in recognition of *Aspergillus spp*: Toll-like receptors (TLRs), C-type lectin receptors (CLRs), and nucleotide oligomerization domain (NOD)–like receptors (NLRs) (Fig. 1). RigI helicases, the fourth PRR family, are not currently known to be involved in *A. fumigatus* recognition.

Toll-like receptors

Both TLR2 and TLR4 signaling are associated with NF-κB translocation and production of proinflammatory cytokines in response to *A. fumigatus in vitro*.[1,2] Initial studies have shown that peritoneal macrophages deficient in TLR2 and MyD88 stimulated with *A. fumigatus* produce significantly lower TNF-α levels, whereas TLR4-deficient peritoneal macrophages produce normal levels of TNF-α when stimulated with the fungus.[3] Blocking TLR4, but not TLR2, resulted in decreased *A. fumigatus*–induced TNF-α production by human adherent monocytes.[4] Furthermore, conidia and hyphae of *A. fumigatus* stimulate cytokine production in macrophages through TLR2, whereas only conidia are able to stimulate macrophages by TLR4.[5] These *in vitro* studies are the first reports that have demonstrated a role for TLRs in the recognition of *A. fumigatus*. Most of these *in vitro* studies controlled for the absence of bacterial LPS.

doi: 10.1111/j.1749-6632.2012.06750.x

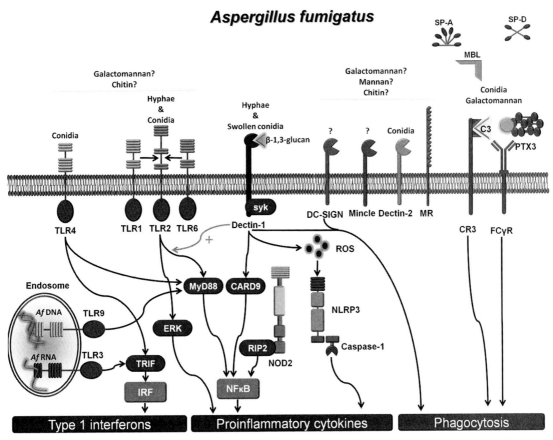

Figure 1. Schematic overview of the PRRs and the downstream signaling pathways involved in the recognition of *A. fumigatus*. Pathogen-associated molecular patterns (PAMPs) of *Aspergillus* lead to various host responses, such as cytokine production and phagocytosis through the activation of PRRs. Syk (spleen tyrosine kinase), DC-SIGN (dendritic cell-specific intercellular adhesion molecule-3-grabbing nonintegrin), MR (mannose receptor), CR3 (complement receptor 3), C3 (complement factor C3), SP-A/D (surfactant protein-A/D), PTX-3 (pentraxin-3), FCγR (FC gamma receptor), *Af* (*Aspergillus fumigatus*), TRIF (TIR-domain–containing adapter-inducing interferon-β), IRF (interferon-regulating factor), ERK (extracellular signal–regulated kinase), MYD88 (myeloid differentiation primary response gene 88), CARD9 (caspase recruitment domain–containing protein 9), RIP2 (receptor interacting protein 2), NF-κB (nuclear factor kappa light chain-enhancer of activated B cells), NOD2 (nucleotide-binding oligomerization domain–containing protein 2), NLRP3 (NACHT, LRR, and PYD domain–containing protein 3), ROS (reactive oxygen species).

In a murine model of IPA, TLR2, and TLR4 deficiency are both correlated with decreased cytokine levels, and TLR2$^{-/-}$, TLR4$^{-/-}$, or MyD88$^{-/-}$ mice have a higher fungal burden than wild-type (WT) mice.[6,7] TLR2 and TLR4 are important for the recruitment of PMNs to the lung during IPA, and the optimal killing of *A. fumigatus in vivo* requires both TLR2 and TLR4.[7] TLR2 can form heterodimers with TLR1 or TLR6, and a recent study has reported that TLR1$^{-/-}$ and TLR6$^{-/-}$ mice have increased fungal burdens during *A. fumigatus* infection. It has been shown that TLR1$^{-/-}$ and TLR6$^{-/-}$ macrophages produce fewer proinflammatory cytokines compared to wild-type mice.[8]

The intracellular TLRs, TLR3, and TLR9 have also been reported to play a role in anti-*Aspergillus* host defense. Epithelial cells can contribute to the protection against *A. fumigatus* in a TLR3-dependent pathway.[9] TLR3 signals through the adaptor molecule TRIF, and TRIF$^{-/-}$ mice are highly susceptible to invasive pulmonary aspergillosis. Furthermore, *A. fumigatus* RNA is recognized by TLR3, and dendritic cells can promote protective cytotoxic T cell responses through a TLR3-mediated pathway.[9] In line with these findings, TLR3$^{-/-}$ mice are highly susceptible to invasive pulmonary aspergillosis.[9] TLR9 can recognize *A. fumigatus* DNA,[10] and TLR9 is actively being recruited to phagosomes that contain *A. fumigatus* conidia.[11] Interestingly, but still largely unexplained, is the finding that TLR9[−/−] mice are protected against lethal *A. fumigatus* infection.[7]

A. fumigatus itself can influence the immune response by modulating TLR signaling. *A. fumigatus* conidia can decrease TLR2 expression on the surface of monocytes by inducing TLR2 internalization, while *A. fumigatus* hyphae can selectively downregulate the TLR4-mediated immune response.[12] Moreover, the *Aspergillus* cell wall components β-glucan and galactomannan suppresses TLR4-induced responses, while α-glucan inhibits TLR2- and TLR4-induced IL-6 production.[13] TLR5 is strongly upregulated by *A. fumigatus* conidia in human monocytes, and although there is no evidence for direct recognition of *A. fumigatus* by TLR5, activation of TLR5 results in decreased capacity of monocytes to inactivate viable *A. fumigatus* conidia.[14] This suggests that the increased expression of TLR5 induced by *A. fumigatus* on the surface of human monocytes can help survival of the fungus when ligands are present that activate TLR5. However, all TLR5 ligands known to date are of bacterial origin, and TLR5 is most likely influencing host defense against *Aspergillus* in the context of cocolonization or coinfection with bacteria.

In line with the *in vitro* and *in vivo* data demonstrating an important role for TLRs in anti-*Aspergillus* host responses, several studies have reported polymorphisms in TLRs that are associated with invasive aspergillosis in patients. A single nucleotide polymorphisms (SNP) in TLR4 that influences TLR4 function is associated with invasive aspergillosis in patients that received hematological stem cell transplantation (HSCT).[15,16] Although, functional studies have demonstrated involvement of TLR2 in the recognition of *A. fumigatus,* to date no polymorphisms in TLR2 have been associated with increased susceptibility to *A. fumigatus* infection. However, SNPs in TLR1 and TLR6, coreceptors that form heterodimers with TLR2, are associated with the development of invasive aspergillosis.[17] A TLR3 SNP has been reported to be associated with increased risk of IPA in a cohort of patients with HSCT.[9]

C-type lectin receptors

The extracellular portion of CLRs consists of several carbohydrate recognition domains, which enable binding to sugar moieties present on microorganisms. Dectin-1 recognizes β-1,3-glucans and is involved in the recognition of *A. fumigatus.*[18] Dectin-1 binding to germ tubes augments TLR2-mediated production of proinflammatory cytokines.[19] Alveolar macrophages induce strong inflammatory responses to swollen and germinating conidia, which correlates with the increased surface expression of β-glucan during germination.[20] The importance of Dectin-1 for anti-*Aspergillus* host defense has been demonstrated by the fact that Dectin-1–deficient mice are more susceptible to invasive pulmonary aspergillosis.[18,21] *A. fumigatus* induces the cytokines IL-17 and IL-22, important for neutrophil recruitment and the production of defensins by epithelial cells, respectively, in a Dectin-1–dependent way.[22–24] It must, however, be noted that the current literature describes both protective and detrimental effects of IL-17 in the host, depending on the host conditions in which the *Aspergillus* infection occurs.[21,24,25] Human airway epithelium and dendritic cells also express Dectin-1, which enables them to recognize *A. fumigatus.*[26] Human monocytes with the Dectin-1 Tyr238X polymorphism that results in an early stop codon produce fewer proinflammatory cytokines in response to *A. fumigatus* stimulation. Importantly, this polymorphism correlates with an increased susceptibility to develop invasive aspergillosis in patients receiving HSCT.[26,27] Recently, a study demonstrated increased susceptibility to invasive aspergillosis in hematological patients that carried SNPs in Dectin-1 intronic regions.[28]

DC-SIGN is a CLR that is expressed on dendritic cells (DCs) and macrophages, and has been shown to be involved in binding and internalization of *A. fumigatus.*[29] Costimulation of neutrophils with *Aspergillus*-stimulated DCs results in the upregulation of costimulatory molecules on DCs, a process that is dependent on DC-SIGN.[30] Notably, SNPs in intronic regions or in the 3′-UTR of the DC-SIGN gene have been associated with increased susceptibility to invasive aspergillosis.[28] Another CLR, the mannose receptor (MR) has been suggested to be involved in the induction of proinflammatory cytokine production induced by *A. fumigatus* conidia that are deficient in melanin.[31] Dectin-2 mediates the release of cysteinyl leukotrienes from murine bone marrow-derived DCs that are stimulated with extracts from *A. fumigatus,* suggesting that Dectin-2 recognizes *A. fumigatus.*[32] However, the function and the importance of the MR and Dectin-2, as well as the role of other CLRs in anti-*Aspergillus* host defense, such as Mincle or macrophage galactose lectin, remains to be determined.

NOD-like receptors

In addition to the TLRs and CLRs, NOD-like receptors (NLRs) that are localized intracellularly have been reported to be involved in the recognition of *Aspergillus*. NLRs are generally subdivided in two categories: NOD1 and NOD2 are the main intracellular receptors for peptidoglycans, whereas the function of the members of the NLRP subfamily is mainly in the formation of the inflammasome. NOD2 is upregulated after exposure to *A. fumigatus*, and recently it was reported that NOD2 recognizes *A. fumigatus*, which subsequently results in the activation of NF-κB.[33] However, individuals completely deficient in NOD2 and suffering from Crohn's disease[34] do not show an increased susceptibility to invasive aspergillosis.

NLRP3 is an NLR that has been extensively studied over recent years. NLRP3 is involved in the activation of the inflammasome, which is a protein complex that can activate caspase-1. Caspase-1 is an enzyme that can cleave the inactive protein pro-IL-1β into the active cytokine IL-1β. *A. fumigatus* hyphae, but not conidia, can activate the NLRP3 inflammasome through the induction of reactive oxygen species (ROS) and potassium efflux.[35] This, in turn, results in the release of active IL-1β. This process is dependent on Syk, which is an adaptor molecule downstream of Dectin-1, suggesting that Dectin-1 is involved in the induction of ROS, the activation of the NLRP3 inflammasome, and release of IL-1β induced by *A. fumigatus*.

Soluble pattern recognition receptors

Collectins are soluble C-type lectins that bind to glycoconjugated structures. Surfactant proteins A (SP-A) and D (SP-D) are collectins that are produced by pneumocytes and that can bind to *A. fumigatus* conidia. The survival of immunosuppressed mice infected with *A. fumigatus* is significantly increased when mice are treated with SP-D.[36] In line with this, SP-D$^{-/-}$ mice die earlier than WT mice, and treatment of SP-D$^{-/-}$ mice with SP-D decreases mortality from 50% to 33%.[37] Interestingly, SP-A$^{-/-}$ mice are less susceptible to invasive as-

pergillosis and SP-A treatment increases mortality in SP-A$^{-/-}$ mice.[37]

Although mannose-binding lectin (MBL) can bind to *A. fumigatus*, this does not necessarily result in enhanced killing of *A. fumigatus*.[38] Immunocompetent mice deficient in MBL are protected against invasive aspergillosis.[39] In contrast, immunosuppressed mice with invasive aspergillosis that are treated with recombinant MBL have a clear survival benefit.[38] Thus, MBL could have different effects depending on the status of the immune system. In addition, low MBL concentrations in serum are associated with higher susceptibility to invasive aspergillosis.[40]

Pentraxin 3 (PTX3) is a soluble pattern recognition receptor that can bind to *Aspergillus* conidia, but not to hyphae.[41] PTX3-deficient neutrophils display impaired phagocytosis of *Aspergillus*.[41] DCs and alveolar macrophages can recognize *A. fumigatus* through a PTX3-dependent pathway, and genetic deletion of PTX3 in mice results in impaired recognition of *A. fumigatus* and increased susceptibility to *A. fumigatus* infection.[41] The complement receptor 3 (CR3) and the FCγR are required for recognition of PTX3-opsonized *Aspergillus*.[42] Notably, NADPH-dependent ROS-deficient mice[43] and corticosteroid immunosuppressed rats,[44] which are both highly susceptible to *Aspergillus* infection, can be rescued by PTX3 treatment. Therefore, PTX3 treatment can prove to be an important immunotherapeutic strategy in invasive aspergillosis.

Pattern recognition receptors orchestrate the immune response

When *A. fumigatus* conidia are inhaled and reach the alveolus, PRRs present on alveolar macrophages and epithelial cells, and the soluble PRRs from the alveolar fluid will recognize the fungus and induce the activation of the immune response (Fig. 2). Alveolar macrophages and epithelial cells will release cytokines and chemokines in response to *A. fumigatus* recruiting innate immune cells, a process dependent on TLR2, TLR4, and Dectin-1.[1–7] Both neutrophils and macrophages recognize and phagocytose *A. fumigatus* conidia by a process that can be enhanced by soluble PRRs such as PTX3 and, possibly, SP-D and MBL;[36,38,42] neutrophils and macrophages can also kill conidia by NADPH-dependent ROS production.[45,46] *A. fumigatus*–induced ROS production can also

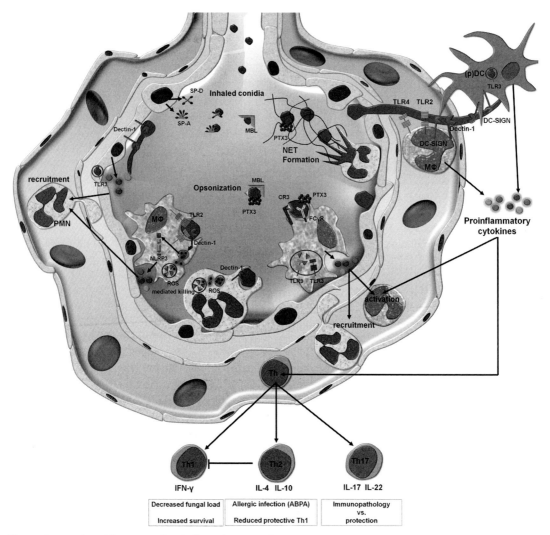

Figure 2. Overview of the PRR-mediated effector functions of innate immune cells that play a role in the pulmonary antifungal host defense against *A. fumigatus*. Inhaled conidia are recognized and opsonized by soluble PRRs-like pentraxin 3 (PTX3), mannose binding lectin (MBL), and surfactant protein-A/D (SP-A and SP-D). Epithelial cells and macrophages (Mφ) produce cytokines through Toll-like receptors (TLRs) that mediate recruitment of innate immune cells like neutrophils (PMNs) and dendritic cells (DCs). Following internalization by dectin-1, phagocytes induce reactive oxygen species (ROS) to mediate eradication of the internalized conidia. Neutrophils generate neutrophil extracellular traps (NETs) by releasing their intracellular contents and DNA. DC-SIGN (dendritic cell–specific intercellular adhesion molecule-3–grabbing nonintegrin), CR3 (complement receptor 3), FCγR (FC gamma receptor), NLRP3 (NACHT, LRR, and PYD domain–containing protein 3).

trigger the NLRP3 inflammasome through Dectin-1 /Syk signaling, resulting in the release of active IL-1β and recruitment of neutrophils.[35] Neutrophils will release neutrophil entrapment traps (NETs) that, in turn, might help to kill and inhibit growth of *A. fumigatus* hyphae that will be too large to be phagocytized.[47,48] Notably, restoration of NADPH oxidase activity by gene therapy in patients with

chronic granulomatous disease provides protection against invasive aspergillosis.[49,50] The restoration of the NADPH oxidase activity restores NET formation and calprotein release, which is associated with more efficient killing of *Aspergillus ex vivo*.[49,50] Recently, plasmocytoid DCs have been reported to have the ability to kill hyphae extracellularly.[51] However, the PRRs involved in this process, and the

PRRs that trigger the formation of NETs in response to *Aspergillus* have not yet been elucidated. If the innate immune system fails to control the infection, antigen-presenting cells such as dendritic cells will activate the adaptive immune response, a process where Dectin-1 and TLR3 play an important role.[9,52]

The interplay between other CLRs and TLR-MyD88 pathways are also important for the induction of the adaptive immune response. Especially Dectin-1, which can regulate the balance between Th1 and Th17 responses.[53] Although TLR-MyD88 signaling is crucial for IFN-γ production in *Aspergillus*-specific CD4 T cells, it is not required for the recruitment or proliferation of these cells in response to *A. fumigatus*.[54] The TLR-MyD88 pathway plays a less important role in the induction of Th2 responses by *A. fumigatus*,[54] in contrast to the induction of Th2 cells by OVA, which is MyD88 dependent.[55]

Concluding remarks

Although PRRs play an important role in the host defense against *A. fumigatus*, it has to be noted that invasive *A. fumigatus* infections only occur in immunosuppressed patients. MyD88$^{-/-}$ mice only develop invasive aspergillosis when they are immunosuppressed, and nonimmunosuppressed patients with MyD88 deficiency are not prone to develop invasive pulmonary aspergillosis.[56,57] Furthermore, humans that lack surface expression of Dectin-1, which is involved in many fundamental anti-*Aspergillus* host responses, do not suffer from invasive aspergillosis when they are not immunosuppressed. These findings suggest that the TLR and Dectin-1 pathways are redundant in anti-*Aspergillus* host defense. On the one hand, this suggests that only concomitant deficiencies in more recognition pathways at the same time will likely result in an increased susceptibility, and this hypothesis is supported by the increased susceptibility to infections in CARD9-deficient patients, who display defects in several CLR pathways, and possible also NOD2-dependent signaling.[58] On the other hand, invasive aspergillosis especially develops in patients that are immunosuppressed, and it is in this particular setting that polymorphisms and deficiencies in TLR, Dectin-1, and other PRR signaling pathways become highly relevant for the antifungal host defense.

Acknowledgments

F.L.vd.V. was supported by a Veni Grant from the Netherlands Organization for Scientific Research. M.G.N. was supported by a Vici Grant from the Netherlands Organization for Scientific Research.

Conflicts of interest

The authors declare no conflicts of interest.

References

1. Meier, A. *et al.* 2003. Toll-like receptor (TLR) 2 and TLR4 are essential for Aspergillus-induced activation of murine macrophages. *Cell Microbiol.* **5:** 561–570.
2. Braedel, S. *et al.* 2004. Aspergillus fumigatus antigens activate innate immune cells via toll-like receptors 2 and 4. *Br. J. Haematol.* **125:** 392–399.
3. Mambula, S.S. *et al.* 2002. Toll-like receptor (TLR) signaling in response to Aspergillus fumigatus. *J. Biol. Chem.* **277:** 39320–39326.
4. Wang, J.E. *et al.* 2001. Involvement of CD14 and toll-like receptors in activation of human monocytes by Aspergillus fumigatus hyphae. *Infect. Immun.* **69:** 2402–2406.
5. Netea, M.G. *et al.* 2003. Aspergillus fumigatus evades immune recognition during germination through loss of toll-like receptor-4-mediated signal transduction. *J. Infect. Dis.* **188:** 320–326.
6. Balloy, V. *et al.* 2005. Involvement of toll-like receptor 2 in experimental invasive pulmonary aspergillosis. *Infect. Immun.* **73:** 5420–5425.
7. Bellocchio, S. *et al.* 2004. The contribution of the Toll-like/IL-1 receptor superfamily to innate and adaptive immunity to fungal pathogens in vivo. *J. Immunol.* **172:** 3059–3069.
8. Rubino, I. *et al.* 2012. Species-specific recognition of Aspergillus fumigatus by Toll-like receptor 1 and Toll-like receptor 6. *J. Infect. Dis.* **205:** 944–954.
9. Carvalho, A. *et al.* 2012. TLR3 essentially promotes protective class I-restricted memory CD8 T-cell responses to Aspergillus fumigatus in hematopoietic transplant patients. *Blood* **119:** 967–977.
10. Ramirez-Ortiz, Z.G. *et al.* 2008. Toll-like receptor 9-dependent immune activation by unmethylated CpG motifs in Aspergillus fumigatus DNA. *Infect. Immun.* **76:** 2123–2129.
11. Kasperkovitz, P.V., M.L. Cardenas & J.M. Vyas. 2010. TLR9 is actively recruited to Aspergillus fumigatus phagosomes and requires the N-terminal proteolytic cleavage domain for proper intracellular trafficking. *J. Immunol.* **185:** 7614–7622.
12. Chai, L.Y. *et al.* 2009. Modulation of Toll-like receptor 2 (TLR2) and TLR4 responses by Aspergillus fumigatus. *Infect. Immun.* **77:** 2184–2192.
13. Chai, L.Y. *et al.* 2011. Aspergillus fumigatus cell wall components differentially modulate host TLR2 and TLR4 responses. *Microbes. Infect.* **13:** 151–159.
14. Rodland, E.K. *et al.* 2011. Toll like receptor 5 (TLR5) may be involved in the immunological response to Aspergillus fumigatus in vitro. *Med. Mycol.* **49:** 375–379.

15. Bochud, P.Y. *et al.* 2008. Toll-like receptor 4 polymorphisms and aspergillosis in stem-cell transplantation. *N. Engl. J. Med.* **359:** 1766–1777.

16. de Boer, M.G. *et al.* 2011. Influence of polymorphisms in innate immunity genes on susceptibility to invasive aspergillosis after stem cell transplantation. *PLoS One* **6:** e18403.

17. Kesh, S. *et al.* 2005. TLR1 and TLR6 polymorphisms are associated with susceptibility to invasive aspergillosis after allogeneic stem cell transplantation. *Ann. N.Y. Acad. Sci.* **1062:** 95–103.

18. Steele, C. *et al.* 2005. The beta-glucan receptor Dectin-1 recognizes specific morphologies of Aspergillus fumigatus. *PLoS Pathog.* **1:** e42.

19. Gersuk, G.M. *et al.* 2006. Dectin-1 and TLRs permit macrophages to distinguish between different Aspergillus fumigatus cellular states. *J. Immunol.* **176:** 3717–3724.

20. Hohl, T.M. *et al.* 2005. Aspergillus fumigatus triggers inflammatory responses by stage-specific beta-glucan display. *PLoS Pathog.* **1:** e30.

21. Werner, J.L. *et al.* 2009. Requisite role for the Dectin-1 beta-glucan receptor in pulmonary defense against Aspergillus fumigatus. *J. Immunol.* **182:** 4938–4946.

22. Chamilos, G. *et al.* 2010. Generation of IL-23 producing dendritic cells (DCs) by airborne fungi regulates fungal pathogenicity via the induction of T(H)-17 responses. *PLoS One* **5:** e12955.

23. Gessner, M.A. *et al.* 2012. Dectin-1-dependent interleukin-22 contributes to early innate lung defense against Aspergillus fumigatus. *Infect. Immun.* **80:** 410–417.

24. Werner, J.L. *et al.* 2011. Neutrophils produce interleukin 17A (IL-17A) in a Dectin-1- and IL-23-dependent manner during invasive fungal infection. *Infect. Immun.* **79:** 3966–3977.

25. Romani, L. *et al.* 2008. Defective tryptophan catabolism underlies inflammation in mouse chronic granulomatous disease. *Nature* **451:** 211–215.

26. Cunha, C. *et al.* 2010. Dectin-1 Y238X polymorphism associates with susceptibility to invasive aspergillosis in hematopoietic transplantation through impairment of both recipient- and donor-dependent mechanisms of antifungal immunity. *Blood* **116:** 5394–5402.

27. Chai, L.Y. *et al.* 2011. The Y238X stop codon polymorphism in the human beta-glucan receptor Dectin-1 and susceptibility to invasive aspergillosis. *J. Infect. Dis.* **203:** 736–743.

28. Sainz, J. *et al.* 2012. Dectin-1 and DC-SIGN polymorphisms associated with invasive pulmonary Aspergillosis infection. *PLoS One* **7:** e32273.

29. Serrano-Gomez, D. *et al.* 2004. Dendritic cell-specific intercellular adhesion molecule 3-grabbing nonintegrin mediates binding and internalization of Aspergillus fumigatus conidia by dendritic cells and macrophages. *J. Immunol.* **173:** 5635–5643.

30. Park, S.J., M.D. Burdick & B. Mehrad. 2012. Neutrophils mediate maturation and efflux of lung dendritic cells in response to Aspergillus germ tubes. *Infect. Immun.* **80:** 1759–1765.

31. Chai, L.Y. *et al.* 2010. Aspergillus fumigatus conidial melanin modulates host cytokine response. *Immunobiology* **215:** 915–920.

32. Barrett, N.A. *et al.* 2009. Dectin-2 recognition of house dust mite triggers cysteinyl leukotriene generation by dendritic cells. *J. Immunol.* **182:** 1119–1128.

33. Li, Z.Z. *et al.* 2012. Role of NOD2 in regulating the immune response to Aspergillus fumigatus. *Inflamm Res.* **61:** 643–648.

34. Ogura, Y. *et al.* 2001. A frameshift mutation in NOD2 associated with susceptibility to Crohn's disease. *Nature* **411:** 603–606.

35. Said-Sadier, N. *et al.* 2010. Aspergillus fumigatus stimulates the NLRP3 inflammasome through a pathway requiring ROS production and the Syk tyrosine kinase. *PLoS One* **5:** e10008.

36. Madan, T. *et al.* 2001. Protective role of lung surfactant protein D in a murine model of invasive pulmonary aspergillosis. *Infect. Immun.* **69:** 2728–2731.

37. Madan, T. *et al.* 2010. Susceptibility of mice genetically deficient in SP-A or SP-D gene to invasive pulmonary aspergillosis. *Mol. Immunol.* **47:** 1923–1930.

38. Kaur, S. *et al.* 2007. Protective role of mannan-binding lectin in a murine model of invasive pulmonary aspergillosis. *Clin. Exp. Immunol.* **148:** 382–389.

39. Clemons, K.V. *et al.* 2010. Resistance of MBL gene-knockout mice to experimental systemic aspergillosis. *Immunol. Lett.* **128:** 105–107.

40. Lambourne, J. *et al.* 2009. Association of mannose-binding lectin deficiency with acute invasive aspergillosis in immunocompromised patients. *Clin. Infect. Dis.* **49:** 1486–1491.

41. Garlanda, C. *et al.* 2002. Non-redundant role of the long pentraxin PTX3 in anti-fungal innate immune response. *Nature* **420:** 182–186.

42. Moalli, F. *et al.* 2010. Role of complement and Fc{gamma} receptors in the protective activity of the long pentraxin PTX3 against Aspergillus fumigatus. *Blood* **116:** 5170–5180.

43. D'Angelo, C. *et al.* 2009. Exogenous pentraxin 3 restores antifungal resistance and restrains inflammation in murine chronic granulomatous disease. *J. Immunol.* **183:** 4609–4618.

44. Lo Giudice, P. *et al.* 2010. Efficacy of PTX3 in a rat model of invasive aspergillosis. *Antimicrob. Agents Chemother.* **54:** 4513–4515.

45. Philippe, B. *et al.* 2003. Killing of Aspergillus fumigatus by alveolar macrophages is mediated by reactive oxidant intermediates. *Infect. Immun.* **71:** 3034–3042.

46. Vethanayagam, R.R. *et al.* 2011. Role of NADPH oxidase versus neutrophil proteases in antimicrobial host defense. *PLoS One* **6:** e28149.

47. Bruns, S. *et al.* 2010. Production of extracellular traps against Aspergillus fumigatus in vitro and in infected lung tissue is dependent on invading neutrophils and influenced by hydrophobin RodA. *PLoS Pathog.* **6:** e1000873.

48. McCormick, A. *et al.* 2010. NETs formed by human neutrophils inhibit growth of the pathogenic mold Aspergillus fumigatus. *Microbes. Infect.* **12:** 928–936.

49. Bianchi, M. *et al.* 2009. Restoration of NET formation by gene therapy in CGD controls aspergillosis. *Blood* **114:** 2619–2622.

50. Bianchi, M. *et al.* 2011. Restoration of anti-Aspergillus defense by neutrophil extracellular traps in human chronic granulomatous disease after gene therapy is calprotectin-dependent. *J. Allergy Clin. Immunol.* **127:** 1243–1252 e1247.

51. Ramirez-Ortiz, Z.G. *et al.* 2011. A nonredundant role for plasmacytoid dendritic cells in host defense against the human fungal pathogen Aspergillus fumigatus. *Cell Host Microbe* **9:** 415–424.

52. Rivera, A. *et al.* 2011. Dectin-1 diversifies Aspergillus fumigatus-specific T cell responses by inhibiting T helper type 1 CD4 T cell differentiation. *J. Exp. Med.* **208:** 369–381.

53. Rivera, A. *et al.* 2011. Dectin-1 diversifies Aspergillus fumigatus-specific T cell responses by inhibiting T helper type 1 CD4 T cell differentiation. *J. Exp. Med.* **208:** 369–381.

54. Rivera, A. *et al.* 2006. Innate immune activation and CD4+ T cell priming during respiratory fungal infection. *Immunity* **25:** 665–675.

55. Piggott, D.A. *et al.* 2005. MyD88-dependent induction of allergic Th2 responses to intranasal antigen. *J. Clin. Invest.* **115:** 459–467.

56. Chignard, M. *et al.* 2007. Role of Toll-like receptors in lung innate defense against invasive aspergillosis. Distinct impact in immunocompetent and immunocompromized hosts. *Clin. Immunol.* **124:** 238–243.

57. von Bernuth, H. *et al.* 2008. Pyogenic bacterial infections in humans with MyD88 deficiency. *Science* **321:** 691–696.

58. Hsu, Y.M. *et al.* 2007. The adaptor protein CARD9 is required for innate immune responses to intracellular pathogens. *Nature Immunol.* **8:** 198–205.

Ann. N.Y. Acad. Sci. ISSN 0077-8923

ANNALS OF THE NEW YORK ACADEMY OF SCIENCES
Issue: *Advances Against Aspergillosis*

Activation of the neutrophil NADPH oxidase by *Aspergillus fumigatus*

Keith B. Boyle,[1] Len R. Stephens,[2] and Phillip T. Hawkins[2]

[1]MRC Laboratory of Molecular Biology, Cambridge, United Kingdom. [2]Inositide Laboratory, Babraham Institute, Babraham, United Kingdom

Address for correspondence: Keith B. Boyle, Division of Protein and Nucleic Acid Chemistry, MRC Laboratory of Molecular Biology, Hills Road, Cambridge, CB2 0QH, United Kingdom. kboyle@mrc-lmb.cam.ac.uk

Upon infection of the respiratory system with the fungus *Aspergillus fumigatus* various leukoctytes, in particular neutrophils, are recruited to the lung to mount an immune response. Neutrophils respond by both phagocytosing conidia and mediating extracellular killing of germinated, invasive hyphae. Of paramount importance to an appropriate immune response is the neutrophil NADPH oxidase enzyme, which mediates the production of various reactive oxygen species (ROS). This is evidenced by the acute sensitivity of both oxidase-deficient humans and mice to invasive aspergillosis. Herein we briefly review the mechanisms and functions of oxidase activation and discuss our recent work identifying at least some of the important players in hyphal-induced oxidase activation and neutrophil function. Among these we define the phosphoinositide 3-kinase enzyme and the regulatory protein Vav to be of critical importance and allude to a kinase-independent role for Syk.

Keywords: reactive oxygen species; phosphoinositide 3-kinase; β_2-integrin; antifungal immunity

Aspergillus fumigatus is a saprophytic fungus whose air-borne spores (conidia) are ubiquitous in the environment and are inhaled daily by humans. Despite its unusualness among fungi in its ability to withstand the relatively high temperatures within the human body, most individuals do not incur noticeable infections during their lifetime. However, certain groups of human patients frequently present with often life-threatening invasive aspergillosis (IA), wherein inhaled conidia germinate and grow into proinflammatory, tissue-damaging hyphae. Such patient groups include those with lymphoma, leukemia or those having undergone bone-marrow transplantation. A common factor among these groups, either as a direct result of the disease itself or as a therapy-induced side effect, is a dramatic fall in the number of neutrophils in the body, known as neutropenia.[1,2]

Neutrophils are key effector cells in the immune response against *A. fumigatus*

Among white-blood cells of the hematopoietic system neutrophils belong to the granulocyte cell family, together with basophils and eosinophils, which are characterized by the presence of granules in their cytoplasm. Neutrophils are the most abundant of these cell types in mammals and are usually the first cell type recruited to sites of infection or injury. Although their life span within the circulatory system may be longer than previously estimated (approximately five days *in vivo* by recent estimates,[3] compared to 6–10 hours *ex vivo*), their numbers can increase up to 10-fold upon infection. The importance of this cell type to an appropriate immune response upon infection with *A. fumigatus* is evident from increased susceptibility of mice treated with neutralizing antibodies against the neutrophil recruiting cytokines MIP-2 or GM-CSF.[4] Moreover, mice that are depleted of neutrophils by means of a specific antibody also display a greatly increased mortality.[5] Among the mechanisms by which neutrophils respond to and eliminate pathogens is the NADPH oxidase Nox2. Upon cell activation this transmembrane enzyme complex mediates the generation of various reactive oxygen species (ROS), some of which are antimicrobidical (more below). Human chronic granulomatous deficiency (CGD) patients, who carry

doi: 10.1111/j.1749-6632.2012.06821.x

inborn mutations in components of the oxidase, frequently present with IA, highlighting the importance of this enzyme in the neutrophil response.[6] Although the NADPH oxidase complex is found in other cell types, such as macrophages, the importance of the oxidase outside of neutrophils with respect to an anti-*Aspergillus* function is controversial, with evidence that oxidase-deficient macrophages can eliminate conidia effectively, at least *ex vivo*.[7]

Neutrophils commonly respond to pathogens by phagocytosing them, followed by maturation of the pathogen-containing phagosome through fusion with their intracellular granules. These granules are classified as primary/azurophilic granules (containing the serine proteases neutrophil elastase and cathepsin G together with myeloperoxidase (MPO)), secondary/specific granules (containing lactoferrin, lysozyme and MPO), and tertiary granules (containing gelatinase).[8] Neutrophils respond to *A. fumigatus* by phagocytosing both dormant conidia and germlings. While dormant conidia appear to be remarkably resistant to killing inside the phagosome, upon swelling and attempted germination they become more susceptible.[9] Hyphal filaments of the fungus are too large to be phagocytosed, so neutrophils attach themselves along the hyphal surface and apparently secrete their granule contents onto the hypha. It has been shown that neutrophils can mediate efficient killing of hyphae by this means.[10] Recent work has demonstrated that neutrophil extracellular traps (NETs) can be elicited by exposure to *A. fumigatus*, though there remains little evidence that this is an effective antifungal mode of action.[11] The fact that CGD patients do not all readily succumb to IA in the face of daily exposure to *A. fumigatus* spores suggests that there may be a means by which neutrophils may deal with conidia in a neutrophil oxidase-indendent manner. For very low inocula it is thought that this may be by means of elimination by macrophages or via lactoferrin-mediated iron sequestration in neutrophils, as a means to prevent conidial growth.[12]

Assembly and activation of the neutrophil NADPH oxidase

In resting neutrophils, the oxidase is partitioned into the membrane-resident flavocytochrome subunits gp91phox and p22phox and the cytosolic components p67phox, p47phox, p40phox, and the small GTP-binding protein Rac. Genetic deletion of either gp91phox or Rac2 has been shown to render mice more susceptible to fungal infection.[7,13] Upon stimulation of the cell, with either soluble agonists or particulate stimuli such as bacteria and fungi, a complex series of events mediates the assembly of the oxidase on a designated membrane. For some time it was thought that this assembly occurred on the plamsa membrane itself or plasma membrane-derived phagosomal membrane, though it is now accepted that assembly occurs on at least some of the intracellular granule membranes.[14] Among the events required for assembly of a functional oxidase are phosphorylation of p47phox on a number of residues, thereby triggering a conformation change in the protein and permitting interaction of p47phox with p22phox.[15] In addition, Rac undergoes activation by means of exchange of the nucleotide GTP for GDP through the action of various guanine nucleotide exchange factor (GEF) proteins, together with loss of binding to its guanine nucleotide dissociation inhibitor (GDI), thereby exposing a lipid-modified membrane-inserting protein tail. Membrane localized GTP-Rac then binds to the TPR domain of p67phox permitting the recruitment of the p67phox/p40phox heterodimer.[15] In addition to these events, critical to the assembly and activation of the oxidase is the generation of specific lipid species in the target membrane.[16] These lipids, collectively termed phosphoinositides, exhibit differential phosphorylation of the various positions of the inositol ring of phosphatidylinositol, and thereby recruit-specific effector proteins. Of importance here is phosphorylation of the dual-phosphorylated lipid $PI(4,5)P_2$ to $PI(3,4,5)P_3$. This latter lipid acts to recruit and activate proteins containing pleckstrin homology (PH) domains, notably certain GEF proteins for Rac. $PI(3,4,5)P_3$ is formed by the action of class I phosphoinositide 3-kinases (PI3Ks), of which there are four members (PI3Kα, β, γ, and δ) expressed in neutrophils.[17,18] Class I PI3Ks are involved in diverse cellular functions with some of those pertinent to neutrophil function including oxidase activation and actin polymerization, the latter being important in the morphological changes required for neutrophil movement toward attractant stimuli or across various substrata.

Once assembled, the oxidase can now accept electrons from NADPH and transfer them across the membrane to molecular oxygen, thereby generating superoxide. The mechanism by which superoxide and its ROS derivatives actually mediate their anti-*Aspergillus* effects has been the subject of controversy.[8] Biochemical experiments over 30 years ago showed that certain ROS, in particular those catalyzed by MPO, such as various hypohalous acids, could damage hyphae.[10] In support of this, MPO-deficient mice exhibit a fairly specific susceptibility to *A. fumigatus* infection.[19] Indeed, exogenous addition of the MPO substrate hydrogen peroxide to oxidase-deficient neutrophils corrects their defect.[20] It had been proposed that the electrogenic generation of ROS on one side of the membrane was coupled to compensatory ion transport and resulting activation of serine proteases such as elastase and cathepsin G.[21] However, neutrophils deficient in the specific ion pump or in the enzyme required for the maturation of serine proteases do not exhibit increased susceptibility to infection.[22,23] Thus, it remains likely that certain ROS species are the main means by which neutrophils damage and kill *A. fumigatus*.

What are the signaling proteins involved in the neutrophil response to hyphae?

In order to mount a destructive ROS response against *A. fumigatus* neutrophils must firstly recognize the pathogen and then coordinate intracellular molecular signaling pathways to assemble a functional oxidase. However, the mechanisms by which neutrophils do this, at least in the case of *A. fumigatus*, are not well understood. Neutrophils express various receptors on their surface, in common with other phagocytic cells, and have been shown to utilize at least some of these for microbial recognition, phagocytosis, and oxidase activation, e.g., Fc receptors and integrins for antibody and complement opsonized targets respectively.[24] However, in contrast to dendritic cells and macrophages, the identity and function of many of the pathogen-related receptors expressed on neutrophils, such as those of the Toll-like receptor family and C-type lectin family, remain understudied. This remains so in part due to the technical difficulty of working with neutrophils because of their short life-span *ex vivo* compared to other immune cell types. Nevertheless, what has been shown is that mice deficient in the cell-surface receptor Dectin-1 are more susceptible to infection with *A. fumigatus* and neutrophils from these mice have an attenuated ROS response to swollen coni-

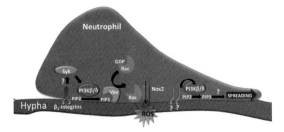

Figure 1. Overview of the signaling mechanisms by which neutrophils respond to hyphae. Neutrophils respond to hyphae by a number of means, including generation of reactive oxygen species (ROS) and spreading along the hyphae so as to maximize the surface area of interaction. The ROS response, mediated by NADPH oxidase (Nox2) activation, is initiated by recognition of hyphae by β_2-integrins on the neutrophil cell surface. Although the precise mechanisms remain unclear this somehow recruits the kinase Syk, which in turn recruits PI3K β and δ isoforms either directly,[38] or indirectly. Activated PI3Ks catalyze the synthesis of PI(3,4,5)P3 (PIP3) from PIP2 in the plasma membrane, which in turn recruits the exchange factor (GEF) Vav, part of whose function is to switch GTP for GDP on Rac, thereby activating it. It has been shown that Vav can interact with Syk directly[39] and also the oxidase subunit p67phox,[40] so all of these proteins may form a large juxta-membrane signaling complex. Activated Rac, together with numerous other events required for oxidase assembly (discussed in the main text), can then aid in activation of the oxidase, resulting in the generation of ROS outside of the neutrophil onto the hyphae. Our observations suggest that the vast majority of ROS produced is directed at the hyphal filament and not simply generated from the entire plasma membrane (data not shown). Concomitant with this are the changes in cell morphology required for spreading of the neutrophil along the hypha. While β_2-integrins may play a minor role here, another as-yet unidentified receptor(s) is required. Nevertheless, this receptor(s) still couples to activation of PI3K β and δ, though the other signaling events that occur remain unknown for now.

dia.[25] Dectin-1 is a receptor for the carbohydrate β-1,3-glucan, which is exposed on conidia only after germination commences with conidial swelling and appears to be masked again in hyphae.[26,27] It is unknown what cellular receptor is required for phagocytosis of dormant conidia and it is unclear as to whether Dectin-1 can recognize the hyphal form. It has been noted that Dectin-1 may only play a minor role in the response to *A. fumigatus* in humans since Dectin-1 deficiency is a susceptibility factor for aspergillosis only in patients already at high risk of infection.[28] Thus, other receptors must be involved. We have attempted to begin to work out the signaling pathways by which neutrophils respond to *A. fumigatus*, at least for the hyphal form against which neutrophils play a unique role. By making use of commercially available lipid and protein kinase inhibitors together with neutrophils purified from genetically targeted mice we have begun to determine which signaling proteins are involved in the neutrophil response to hyphae. We found that lipid products of class I PI3Ks were generated at the neutrophil membrane in contact with the hypal filament, that PI3K β and δ subunits played overlapping, partially redundant roles in not only oxidase activation but also in the morphological changes required for spreading of the neutrophil along the hyphal surface. In addition, the PI3K effector protein Vav (a GEF for Rac) and the kinase Syk were essential for oxidase activation (Fig. 1). Upstream of these components it was found, somewhat surprisingly, that Dectin-1 only played a very minor role in neutrophil responses. However, neutrophils from mice deficient in the multifunctional β_2-integrin receptor family had markedly impaired oxidase activation, though not neutrophil spreading.[29] It is likely that it is the β_2-integrin isoform Mac-1 that is important in this respect, as previous work has shown that this isoform can recognize various glucose-based carbohydrates.[30] However, the identity of the ligand(s) for integrins on hyphae remains unknown. There were two findings within this work that for now remain unexplained and require further attention. Firstly, what is the precise molecular signaling that links β_2-integrins specifically to oxidase activation and, secondly, what are the other neutrophil receptors involved, in particular for PI3K activation?

Recent work has shown that β_2-integrins couple to downstream oxidase activation in neutrophils via two adaptor proteins, FcRγ and DAP12, that contain a classical Immunoreceptor tyrosine-based activation motif (ITAM).[31,32] Phosphorylation of these adaptor proteins recruits the kinase Syk, a protein that plays important roles in various aspects of the hematopoietic system.[33] Syk, in turn, phosphorylates a number of target proteins, including itself, to propagate the appropriate signals. However, although deletion of Syk abolished a neutrophil ROS response to hyphae, we found that deletion of both FcRγ and DAP12 had no effect on oxidase activation.[29] Furthermore, selective inhibition of Syk activity with a small molecule inhibitor was much less potent at inhibiting hyphal-induced ROS compared to immune complex-stimulated ROS (a known Syk-dependent response; see Fig. 2). To our knowledge this is the first hint of a kinase-independent role for

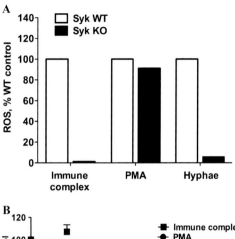

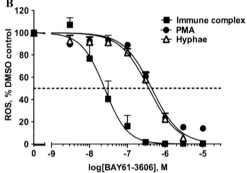

Figure 2. Syk protein but not kinase activity is required for hyphal-induced oxidase activation. Syk is a tyrosine kinase involved in relaying signals from receptors to the cell interior for various cell types of the hematopoietic system. Its importance is demonstrated by the embryonic lethality of mice deficient for the protein.[41] The specific involvement of Syk in hematopoietic cellular functions of interest can be interrogated by means of reconstitution of the hematopoietic system of lethally irradiated mice using stem cells from genetically deficient fetal livers.[42] This approach has shown that Syk is required for functional responses downstream of various receptors on neutrophils, such as antibody Fc receptors, certain C-type lectins, and β_2-integrins.[33] (A) We found that Syk-deficient neutrophils purified from reconstituted mice had a severely impaired ROS response to both a synthetic antigen-antibody immune complex, as expected, and hyphae. A Syk-independent stimulus (PMA) was unaffected by Syk deletion, evidence that a functional oxidase complex can be assembled (data are from a single experiment, representative of two). (B) The catalytic activity of Syk was investigated by means of a Syk-selective inhibitor (BAY61–3606) whose IC_{50} *in vitro* is around 7.5 nM.[43] Neutrophils were titrated with increasing concentrations of inhibitor and the ROS response to the same three stimuli measured by means of a luminol chemiluminescence assay (as in Boyle *et al.*). An inhibitory curve with an IC_{50} of approximately 25 nM was observed for the immune complex positive control. In contrast, an IC_{50} of 360 nM was observed for hyphae, similar to that for the Syk-independent stimulus PMA (430 nM), indicating that these are likely off-target effects of the inhibitor at these concentrations. (Error bars represent the SEM of three indepenent experiments, with neutrophils from two mice combined per experiment.)

Syk. Thus, either a novel paradigm for β_2-integrin-induced signaling exists or another, as yet unidentified, adaptor protein or integrin-coupled coreceptor is involved. The former may involve direct interaction of Syk with β_2-integrins, a concept shown thus far only in transfected cell lines,[34] whereas the latter is supported by our observation that, unlike oxidase activation, both the neutrophil spreading response and PI3K activation are refractory to integrin deletion, indicating that other receptor(s) are involved. Interestingly, preincubation of neutrophils with a soluble glucan preparation, in the form of laminarin, potently inhibited PI3K activation suggesting that there may be a non-Dectin-1 glucan receptor involved,[29] for example one for β-1,6-glucan, a potent neutrophil stimulus.[35]

While our own work has made some inroads into establishing what events occur inside neutrophils as they mount their response to *Aspergillus* much further work is clearly required to determine the precise signaling that occurs and establish the mechanisms by which neutrophils kill this fungus. It should not be forgotten that *A. fumigatus* has coevolved with its host organisms (thought to not necessarily be humans in this context but various water-fowl) over a significant period of time and as a result has evolved strategies to counteract the host immune response. For example, the secreted toxins gliotoxin and fumagillin exhibit antagonistic effects against neutrophil oxidase activation.[36,37] Thus, future studies investigating the immune response of neutrophils against the fungus should bear this in mind.

Conflicts of interest

The authors declare no conflicts of interest.

References

1. Hasenberg, M. *et al.* 2011. Phagocyte responses towards Aspergillus fumigatus. *Int. J. Med. Microbiol.* **301:** 436–444.
2. Lin, S.J., J. Schranz & S.M. Teutsch. 2001. Aspergillosis case-fatality rate: systematic review of the literature. *Clin. Infect. Dis.* **32:** 358–366.
3. Pillay, J. *et al.* 2010. In vivo labeling with 2H2O reveals a human neutrophil lifespan of 5.4 days. *Blood* **116:** 625–627.
4. Schelenz, S., D.A. Smith & G.J. Bancroft. 1999. Cytokine and chemokine responses following pulmonary challenge with Aspergillus fumigatus: obligatory role of TNF-alpha and GM-CSF in neutrophil recruitment. *Med. Mycol.* **37:** 183–194.
5. Mircescu, M.M. *et al.* 2009. Essential role for neutrophils but not alveolar macrophages at early time points following Aspergillus fumigatus infection. *J. Infect. Dis.* **200:** 647–656.
6. Kang, E.M. *et al.* 2010. Retrovirus gene therapy for X-linked chronic granulomatous disease can achieve stable long-term correction of oxidase activity in peripheral blood neutrophils. *Blood* **115:** 783–791.
7. Morgenstern, D.E. *et al.* 1997. Absence of respiratory burst in X-linked chronic granulomatous disease mice leads to abnormalities in both host defense and inflammatory response to Aspergillus fumigatus. *J. Exp. Med.* **185:** 207–218.
8. Nauseef, W.M. 2007. How human neutrophils kill and degrade microbes: an integrated view. *Immunol. Rev.* **219:** 88–102.
9. Levitz, S.M. & R.D. Diamond. 1985. Mechanisms of resistance of Aspergillus fumigatus Conidia to killing by neutrophils in vitro. *J. Infect. Dis.* **152:** 33–42.
10. Diamond, R.D. & R .A. Clark. 1982. Damage to Aspergillus fumigatus and Rhizopus oryzae hyphae by oxidative and nonoxidative microbicidal products of human neutrophils in vitro. *Infect. Immun.* **38:** 487–495.
11. Bruns, S. *et al.* 2010. Production of extracellular traps against Aspergillus fumigatus in vitro and in infected lung tissue is dependent on invading neutrophils and influenced by hydrophobin RodA. *PLoS Pathog.* **6:** e1000873.
12. Zarember, K.A. *et al.* 2007. Human polymorphonuclear leukocytes inhibit Aspergillus fumigatus conidial growth by lactoferrin-mediated iron depletion. *J. Immunol.* **178:** 6367–6373.
13. Roberts, A.W. *et al.* 1999. Deficiency of the hematopoietic cell-specific Rho family GTPase Rac2 is characterized by abnormalities in neutrophil function and host defense. *Immunity* **10:** 183–196.
14. Karlsson, A. & C. Dahlgren. 2002. Assembly and activation of the neutrophil NADPH oxidase in granule membranes. *Antioxid. Redox. Signal* **4:** 49–60.
15. Groemping, Y. & K. Rittinger. 2005. Activation and assembly of the NADPH oxidase: a structural perspective. *Biochem. J.* **386:** 401–416.
16. Perisic, O. *et al.* 2004. The role of phosphoinositides and phosphorylation in regulation of NADPH oxidase. *Adv. Enzyme. Regul.* **44:** 279–298.
17. Hawkins, P.T. *et al.* 2006. Signaling through Class I PI3Ks in mammalian cells. *Biochem. Soc. Trans.* **34:** 647–662.
18. Vanhaesebroeck, B. *et al.* 2010. The emerging mechanisms of isoform-specific PI3K signaling. *Nat. Rev. Mol. Cell Biol.* **11:** 329–341.
19. Aratani, Y. *et al.* 2002. Relative contributions of myeloperoxidase and NADPH-oxidase to the early host defense against pulmonary infections with Candida albicans and Aspergillus fumigatus. *Med. Mycol.* **40:** 557–563.
20. Washburn, R.G., J.I. Gallin & J.E. Bennett. 1987. Oxidative killing of Aspergillus fumigatus proceeds by parallel myeloperoxidase-dependent and -independent pathways. *Infect. Immun.* **55:** 2088–2092.
21. Reeves, E.P. *et al.* 2002. Killing activity of neutrophils is mediated through activation of proteases by K+ flux. *Nature* **416:** 291–297.
22. Vethanayagam, R.R. *et al.* 2011. Role of NADPH oxidase versus neutrophil proteases in antimicrobial host defense. *PLoS One* **6:** e28149.
23. Essin, K. *et al.* 2007. Large-conductance calcium-activated potassium channel activity is absent in human and mouse

neutrophils and is not required for innate immunity. *Am J. Physiol. Cell Physiol.* **293:** C45–C54.

24. Abram, C.L. & C.A. Lowell. 2007. Convergence of immunoreceptor and integrin signaling. *Immunol. Rev.* **218:** 29–44.

25. Werner, J.L. *et al.* 2009. Requisite role for the dectin-1 beta-glucan receptor in pulmonary defense against Aspergillus fumigatus. *J. Immunol.* **182:** 4938–4946.

26. Steele, C. *et al.* 2005. The beta-glucan receptor dectin-1 recognizes specific morphologies of Aspergillus fumigatus. *PLoS Pathog.* **1:** e42.

27. Hohl, T.M. *et al.* 2005. Aspergillus fumigatus triggers inflammatory responses by stage-specific beta-glucan display. *PLoS Pathog.* **1:** e30.

28. Cunha, C. *et al.* 2010. Dectin-1 Y238X polymorphism associates with susceptibility to invasive aspergillosis in hematopoietic transplantation through impairment of both recipient- and donor-dependent mechanisms of antifungal immunity. *Blood* **116:** 5394–5402.

29. Boyle, K.B. *et al.* 2011. Class IA phosphoinositide 3-kinase beta and delta regulate neutrophil oxidase activation in response to Aspergillus fumigatus hyphae. *J. Immunol.* **186:** 2978–2989.

30. Thornton, B.P. *et al.* 1996. Analysis of the sugar specificity and molecular location of the beta-glucan-binding lectin site of complement receptor type 3 (CD11b/CD18). *J. Immunol.* **156:** 1235–1246.

31. Mocsai, A. *et al.* 2002. Syk is required for integrin signaling in neutrophils. *Immunity* **16:** 547–558.

32. Jakus, Z. *et al.* 2007. Immunoreceptor-like signaling by beta 2 and beta 3 integrins. *Trends Cell Biol.* **17:** 493–501.

33. Mocsai, A., J. Ruland & V.L. Tybulewicz. 2010. The SYK tyrosine kinase: a crucial player in diverse biological functions. *Nat. Rev. Immunol.* **10:** 387–402.

34. Woodside, D.G. *et al.* 2001. Activation of Syk protein tyrosine kinase through interaction with integrin beta cytoplasmic domains. *Curr. Biol.* **11:** 1799–1804.

35. Rubin-Bejerano, I. *et al.* 2007. Phagocytosis by human neutrophils is stimulated by a unique fungal cell wall component. *Cell Host Microbe.* **2:** 55–67.

36. Tsunawaki, S. *et al.* 2004. Fungal metabolite gliotoxin inhibits assembly of the human respiratory burst NADPH oxidase. *Infect. Immun.* **72:** 3373–3382.

37. Fallon, J.P., E.P. Reeves & K. Kavanagh. 2010. Inhibition of neutrophil function following exposure to the Aspergillus fumigatus toxin fumagillin. *J. Med. Microbiol.* **59:** 625–633.

38. Moon, K.D. *et al.* 2005. Molecular basis for a direct interaction between the Syk protein-tyrosine kinase and phosphoinositide 3-kinase. *J. Biol. Chem.* **280:** 1543–1551.

39. Deckert, M. *et al.* 1996. Functional and physical interactions of Syk family kinases with the Vav proto-oncogene product. *Immunity* **5:** 591–604.

40. Ming, W. *et al.* 2007. The Rac effector p67phox regulates phagocyte NADPH oxidase by stimulating Vav1 guanine nucleotide exchange activity. *Mol. Cell Biol.* **27:** 312–323.

41. Turner, M. *et al.* 1995. Perinatal lethality and blocked B-cell development in mice lacking the tyrosine kinase Syk. *Nature* **378:** 298–302.

42. Kiefer, F. *et al.* 1998. The Syk protein tyrosine kinase is essential for Fcgamma receptor signaling in macrophages and neutrophils. *Mol. Cell Biol.* **18:** 4209–4220.

43. Yamamoto, N. *et al.* 2003. The orally available spleen tyrosine kinase inhibitor 2-[7-(3,4-dimethoxyphenyl)-imidazo[1,2-c]pyrimidin-5-ylamino]nicotinamide dihydrochloride (BAY 61–3606) blocks antigen-induced airway inflammation in rodents. *J. Pharmacol Exp. Ther.* **306:** 1174–1181.

Ann. N.Y. Acad. Sci. ISSN 0077-8923

ANNALS OF THE NEW YORK ACADEMY OF SCIENCES

Issue: *Advances Against Aspergillosis*

The top three areas of basic research on *Aspergillus fumigatus* in 2011

Nir Osherov

Department of Clinical Microbiology and Immunology, Sackler School of Medicine Ramat-Aviv, Tel-Aviv, Israel

Address for correspondence: Nir Osherov, Ph.D., Department of Clinical Microbiology and Immunology, Sackler School of Medicine, Tel-Aviv University, Ramat-Aviv, Tel-Aviv, Israel. nosherov@post.tau.ac.il

Over 450 peer-reviewed papers containing the keyword *Aspergillus fumigatus* were published in 2011. Although this method may be an impossible task, I have selected three clusters of papers describing exciting recent advances in research on *A. fumigatus*. The first is the novel approach of *in vivo* imagining of experimental aspergillosis by the use of ^{68}Ga-labeled siderophores, internalized by the fungus, and detected via positron emission tomography to image the site infection. This work may lead to improved diagnosis of aspergillosis. The second important finding is that NK lymphocytes, not thought to be involved in host resistance to aspergillosis, can kill aspergilli through direct contact, either through perforin or interferon-γ, or both. The third area pertains to a novel first-in-class antifungal drug, E1210 (Eisai), which inhibits GPI anchoring of fungal-associated cell wall proteins. Thus far, it shows promising *in vitro* activity against a broad range of fungi including Aspergilli, as well as those that are difficult to treat with currently available therapies. Overall, these three areas demonstrate the exciting promise, progress, and utility of basic research against *A. fumigatus*.

Keywords: fumigatus; antifungal; NK cell; imaging

Aspergillus fumigatus is the most common op-portunistic mold pathogen found in humans, causing invasive diseases in immunocompromised patients.[1] The explosive fourfold increase in the incidence of invasive pulmonary aspergillosis (IPA) that has occurred over the last 30 years has triggered a parallel 400% increase in the number of scientific research papers devoted to studying *A. fumigatus*. Highlighting this increased interest, three of these papers have been published in the prestigious publications *Nature* and *Science* over the last five years alone.[2–4] Interestingly, approximately two-thirds of the publications on *A. fumigatus* in 2011 focused on basic scientific research of this mold, while only one-third dealt with clinical aspects, suggesting that we are still trying to gain a better molecular understanding of this pathogen, before we can manipulate its weaknesses to improve treatment.

In this review, I highlight three outstanding basic science research papers published in 2011, which mark the way for future advances in basic and clinical research of invasive aspergillosis.

Imaging IPA by hijacking the iron uptake system of the fungus

Developing imaging modalities that are able to detect IPA with high sensitivity and specificity should allow the early initiation of appropriate antifungal treatment in high-risk patients. Here, Decristo-foro *et al.* from the Innsbruck Medical University in Austria have used a novel approach to this challenging problem. During lung infection, *A. fumigatus* encounters an essentially iron-free environment. All available iron is tightly bound by host chelators, and in particular the protein transferrin. However, in the lungs, *A. fumigatus* can acquire iron by activating two independent high-affinity iron-uptake systems.[5] These originally evolved to enable *A. fumigatus* to overcome iron shortages in its natural soil environment. The most important of the iron uptake systems uses siderophores, small secreted molecules with very high binding affinity to ferric (Fe^{3+}) iron. During infection, *A. fumigatus* upregulates gene clusters involved in the biosynthesis of the

doi: 10.1111/j.1749-6632.2012.06798.x

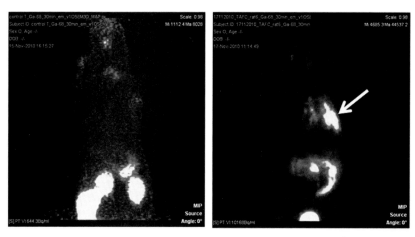

Figure 1. Micro-PET image (maximum intensity projection) of two rats, 1 h postinjection of ^{68}Ga-Triacetylfusarinine (TAFC); left: control uninfected rat, activity is seen in kidney and bladder (urinary excretion); right: rat with severe invasive pulmonary aspergillosis (two days after intrapulmonary instillation of *A. fumigatus* conidia) showing additional high uptake in the left, infected lung (arrow). (Courtesy of C. Decristoforo, Department of Nuclear Medicine, Innsbruck Medical University, Austria; and P. Laverman, Department of Nuclear Medicine, Radboud University Nijmegen Medical Center, the Netherlands.)

siderophores fusarinine (FsC) and triacetylfusarinine (TAFC).[6] Once synthesized and secreted, TAFC and FsC sequester iron even when it is bound to human transferrin, enabling fungal infection to proceed. *A. fumigatus* mutant strains lacking the ability to synthesize FC and TAFC are completely avirulent in mice.[7] Siderophore uptake in *A. fumigatus* is actively mediated by Siderophore Iron Transporters (SIT), permeases of the major facilitator superfamily.[5] Importantly, mammalian cells lack SIT homologs and cannot actively take up siderophores. In two recent papers,[8,9] Petrik *et al.* took advantage of this distinction and designed various purified siderophores in which chelated iron was replaced by radioactive gallium-68 (^{68}Ga). This element has a similar charge and size to iron and is widely used for imaging by positron emission tomography (PET). Once injected into rats infected intratracheally with *A. fumigatus*, the labeled siderophores were actively and selectively concentrated into the infecting fungus, enabling it to be clearly visualized inside the lungs by PET (Fig. 1). In essence, the researchers exploited the same siderophores used so successfully by the fungus to survive during infection, labeled with ^{68}Ga, as Trojan horses to allow specific imaging of the infectious process. Of the ^{68}Ga-labeled siderophores tested, *A. fumigatus* TAFC and bacterial ferrioxamine E were shown to be the best siderophore for clearly and selectively imaging *A. fumigatus* infection. Labeling was sensitive

enough to differentiate between severe and mild infection as early as three days after administration of the fungus.[8] Other fungi and some bacteria are able to actively take up TAFC or ferrioxamine E. Consequently, although further research will be necessary to address the sensitivity and specificity of this method versus existing PCR and ELISA technologies, it may be possible in the near future to use this technology to accurately detect and locate invasive fungal infection at its earliest onset.

NK cells, a novel line of defense against IPA

To date, it has been widely accepted that the major known lines of innate immune defense active against IPA are those mediated by macrophages, neutrophils and dendritic cells.[10] Here, two groups, led by Juergen Loeffler from Wurzburg University and by Thomas Lehrnbecher from Goethe University, Germany, suggest an additional cellular player: the NK (natural killer) cell.[11,12] NK cells are a third class of lymphocyte, related to B and T cells. First recognized for their ability to autonomously recognize and destroy cancer cells, they are now known to participate in the defense against viruses, bacteria, and protozoans.[13] Their response to pathogens generally requires signals (both contact dependent and soluble) from accessory cells such as dendritic cells and macrophages, and involves the release of interferon γ (IFN-γ) and direct cytolytic killing.

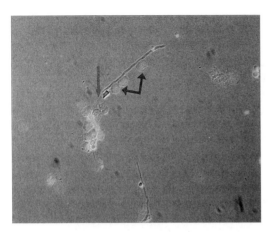

Figure 2. NK cell–*A. fumigatus* interaction is mediated by direct physical contact. NK cells were stained with a DAPI dye (blue arrows), and *A. fumigatus* hyphae were stained with an FITC dye (green arrow) displaying a direct contact after 3 h of coincubation. The photo was acquired using a Zeiss fluorescence microscope, and Zeiss AxioVision LE software (version 4.7) at a magnification of 40×. (Courtesy of J. Loeffler, Universitatsklinikum Wurzburg, Wurzburg, Germany.)

Previous studies on the role of NK cells in IPA suggested that they are the main source of early IFN-γ in the infected lungs, and this is an important mechanism in the defense against this infection.[14,15] Now, Schmidt *et al.*[12] and Bouzani *et al.*[11] demonstrate that purified human NK cells directly recognize and destroy growing *A. fumigatus* in the absence of accessory cells (Fig. 2). Interestingly, although NK cell activation depended on direct contact with the fungus, killing was mediated by a NK cell–secreted factor, which one group identified as perforin (a secreted pore-forming protein) while the other identified as IFN-γ.[11,12] The reason for this disparity remains unclear and will require further investigation. However, both studies point toward NK cells as a potentially important and hitherto unappreciated player in the innate defense against *A. fumigatus* and as a promising new avenue of immunotherapeutic augmentation.

E1210: A novel antifungal that targets anchoring of cell wall proteins

The cell wall is an essential component of all fungi. Since it is not found in mammalian cells, it presents an attractive drug target. Surprisingly, however, of the four major existing families of antifungals (polyenes, azoles, allylamines, and echinocandins) only the latter directly targets the cell wall, by in-

hibiting the enzyme glucan synthase, which is responsible for synthesizing β-1,3-glucan, a major polysaccharide wall component.[16] Recently, however, Asada *et al.* from the Eisai Company in Japan have published a stream of seven papers describing the antifungal activity of a novel cell wall–targeting drug E1210.[17–23] Unlike the echinocandins, which target synthesis of the cell wall polysaccharide scaffold, E1210 inhibits an early step in the glycosylphosphatidylinositol (GPI)-dependent anchoring of cell wall proteins within this scaffold. Lacking these proteins, the cell wall weakens, resulting in the fungistatic arrest of growth. The GPI anchor is synthesized and attached to target proteins within the endoplasmatic reticulum (ER) in a pathway containing approximately 11 enzymes.[24] The target of E1210 is Gwp1p, the fourth enzyme in the pathway, responsible for inositol acylation.[23,25,26] Although mammals also contain the Gwp1 homolog PigW, it is only 28% identical to the fungal gene and is not inhibited by E1210.[23] *In vitro*, E1210 was highly effective (MIC$_{90}$ ranges of 5–200 ng/mL range) against most fungi, including yeast (*Candida* species except *C. krusei*) and molds (*Aspergillus* species, *Fusarium* spp., black molds), as well as strains resistant to azoles and polyenes.[19–22] E1210 was moderately effective against species of zygomycetes (MIC$_{90}$ ranges of 1–8 μg/mL). *In vivo*, E1210 was effective (>80% two-week survival, 2.5–25 mg/kg/day) in the treatment of murine models of disseminated candidiasis (*C. albicans*), pulmonary aspergillosis (*A. fumigatus* or *A. flavus*), and disseminated fusariosis (*Fusarium solani*). E1210 was nontoxic at 100 mg/kg and was generally well tolerated at doses of up to 300 mg/kg in rats.[18] Currently pharmacodynamic and metabolic studies of E1210 are being conducted in rats, dogs, and monkeys, and these will hopefully pave the way for future clinical studies with this novel acting and promising compound leading to improved therapies especially for organisms resistant to currently available treatments.

Acknowledgments

I would like to thank Dr. Ronen Ben Ami for critical reading of this manuscript.

Conflicts of interest

The author declares no conflicts of interest.

References

1. Ben-Ami, R., R.E. Lewis & D.P. Kontoyiannis. 2010. Enemy of the (immunosuppressed) state: an update on the pathogenesis of *Aspergillus fumigatus* infection. *Br. J. Haematol.* **150:** 406–417.

2. Bayram, O. *et al.* 2008. *VelB/VeA/LaeA* complex coordinates light signal with fungal development and secondary metabolism. *Science* **320:** 1504–1506.

3. Aimanianda, V. *et al.* 2009. Surface hydrophobin prevents immune recognition of airborne fungal spores. *Nature* **460:** 1117–1121.

4. O'Gorman, C.M., H.T. Fuller & P.S. Dyer. 2009. Discovery of a sexual cycle in the opportunistic fungal pathogen *Aspergillus fumigatus*. *Nature* **457:** 471–474.

5. Schrettl, M. & H. Haas. 2011. Iron homeostasis—Achilles' heel of *Aspergillus fumigatus*? *Curr. Opin. Microbiol.* **14:** 400–405.

6. McDonagh, A. *et al.* 2008. Sub-telomere directed gene expression during initiation of invasive aspergillosis. *PLoS Pathog.* **4:** e1000154.

7. Schrettl, M. *et al.* 2004. Siderophore biosynthesis but not reductive iron assimilation is essential for *Aspergillus fumigatus* virulence. *J. Exp. Med.* **200:** 1213–1219.

8. Petrik, M. *et al.* 2010. ^{68}Ga-siderophores for PET imaging of invasive pulmonary aspergillosis: proof of principle. *J. Nucl. Med.* **51:** 639–645.

9. Petrik, M. *et al.* 2012. In vitro and in vivo evaluation of selected $^{(68)}$Ga-siderophores for infection imaging. *Nucl. Med. Biol.* **39:** 361–369

10. Park, S.J. & B. Mehrad. 2009. Innate immunity to Aspergillus species. *Clin. Microbiol. Rev.* **22:** 535–551.

11. Bouzani, M. *et al.* 2011. Human NK cells display important antifungal activity against *Aspergillus fumigatus*, which is directly mediated by IFN-gamma release. *J. Immunol.* **187:** 1369–1376.

12. Schmidt, S. *et al.* 2010. Human natural killer cells exhibit direct activity against *Aspergillus fumigatus* hyphae, but not against resting conidia. *J. Infect. Dis.* **203:** 430–435.

13. Newman, K.C. & E.M. Riley. 2007. Whatever turns you on: accessory-cell-dependent activation of NK cells by pathogens. *Nat. Rev. Immunol.* **7:** 279–291.

14. Morrison, B.E. *et al.* 2003. Chemokine-mediated recruitment of NK cells is a critical host defense mechanism in invasive aspergillosis. J. Clin. Invest. **112:** 1862–1870.

15. Park, S.J. *et al.* 2009. Early NK cell-derived IFN-{gamma} is essential to host defense in neutropenic invasive aspergillosis. *J. Immunol.* **182:** 4306–4312.

16. Pitman, S.K., R.H. Drew & J.R. Perfect. 2011. Addressing current medical needs in invasive fungal infection prevention and treatment with new antifungal agents, strategies and formulations. *Expert Opin. Emerg. Drugs* **16:** 559–586.

17. Castanheira, M. *et al.* 2011. Activities of E1210 and comparator agents tested by CLSI and EUCAST broth microdilution methods against Fusarium and Scedosporium species identified using molecular methods. *Antimicrob. Agents Chemother.* **56:** 352–357.

18. Hata, K. *et al.* 2011. Efficacy of oral E1210, a new broad-spectrum antifungal with a novel mechanism of action, in murine models of candidiasis, aspergillosis, and fusariosis. *Antimicrob. Agents Chemother.* **55:** 4543–4551.

19. Miyazaki, M. *et al.* 2011. In vitro activity of E1210, a novel antifungal, against clinically important yeasts and molds. *Antimicrob. Agents Chemother.* **55:** 4652–4658.

20. Pfaller, M.A. *et al.* 2011. In vitro activity of a novel broad-spectrum antifungal, E1210, tested against *Aspergillus spp.* determined by CLSI and EUCAST broth microdilution methods. *Antimicrob. Agents Chemother.* **55:** 5155–5158.

21. Pfaller, M.A. *et al.* 2011. In vitro activity of a novel broad-spectrum antifungal, E1210, tested against Candida spp. as determined by CLSI broth microdilution method. Diagn. Microbiol. Infect. Dis. **71:** 167–170.

22. Pfaller, M.A. *et al.* 2011. Pre-clinical development of antifungal susceptibility test methods for the testing of the novel antifungal agent E1210 versus Candida: comparison of CLSI and European Committee on Antimicrobial Susceptibility Testing methods. *J. Antimicrob. Chemother.* **66:** 2581–2584.

23. Watanabe, N.A. *et al.* 2011. E1210, a new broad-spectrum antifungal, suppresses candida albicans hyphal growth through inhibition of glycosylphosphatidylinositol biosynthesis. *Antimicrob. Agents Chemother.* **56:** 960–971.

24. Orlean, P. & A.K. Menon. 2007. Thematic review series: lipid posttranslational modifications. GPI anchoring of protein in yeast and mammalian cells, or: how we learned to stop worrying and love glycophospholipids. *J. Lipid Res.* **48:** 993–1011.

25. Umemura, M. *et al.* 2003. *GWT1* gene is required for inositol acylation of glycosylphosphatidylinositol anchors in yeast. *J. Biol. Chem.* **278:** 23639–23647.

26. Tsukahara, K. *et al.* 2003. Medicinal genetics approach towards identifying the molecular target of a novel inhibitor of fungal cell wall assembly. *Mol. Microbiol.* **48:** 1029–1042.

Ann. N.Y. Acad. Sci. ISSN 0077-8923

ANNALS OF THE NEW YORK ACADEMY OF SCIENCES
Issue: *Advances Against Aspergillosis*

The cell biology of the innate immune response to *Aspergillus fumigatus*

Michael K. Mansour,[1,2] Jenny M. Tam,[1,2] and Jatin M. Vyas[1,2,3]

[1]Division of Infectious Disease, Department of Medicine, Massachusetts General Hospital, Boston, Massachusetts. [2]Department of Medicine, Harvard Medical School, Boston, Massachusetts. [3]Program in Immunology, Harvard Medical School, Boston, Massachusetts

Address for correspondence: Jatin M. Vyas, M.D., Ph.D., Massachusetts General Hospital, Division of Infectious Disease, 55 Fruit Street, Gray-Jackson Building, Room 504, Boston, MA 02114. jvyas@partners.org

The development of invasive aspergillosis is a feared complication for immunocompromised patients. Despite the use of antifungal agents with excellent bioactivity, the morbidity and mortality rates for invasive aspergillosis remain unacceptably high. Defects within the innate immune response portend the highest risk for patients, but detailed knowledge of molecular pathways in neutrophils and macrophages in response to this fungal pathogen is lacking. Phagocytosis of fungal spores is a key step that places the pathogen into a phagosome, a membrane-delimited compartment that undergoes maturation and ultimately delivers antigenic material to the class II MHC pathway. We review the role of Toll-like receptor 9 (TLR9) in phagosome maturation of *Aspergillus fumigates*–containing phagosomes. Advanced imaging modalities and the development of fungal-like particles are promising tools that will aid in the dissection of the molecular mechanism to fungal immunity.

Keywords: innate immunity; Dectin-1; Toll-like receptor; phagosome; macrophage; TLR9

Introduction

For breath is life, and if you breathe well you will live long on earth.

—Sanskrit proverb

Exposure to *Aspergillus fumigatus* is a universal phenomenon. In immunocompetent individuals, inhalation of these fungal spores is taken up by neutrophils and macrophages that quickly neutralize any potential invasion or infection.[1] However, through the advances of medical care, a growing number of patients have acquired immunological defects either by cellular deficiencies/dysfunction or administration of therapies that modulates the immune response.[2] These immunologically "fragile" patients possess a higher risk for development of clinically significant invasive fungal infections, including aspergillosis. Indeed, over 10% of patients who receive allogeneic bone marrow transplantation develop invasive aspergillosis during their clinical course.[3] Antimicrobial therapies that have excellent *in vitro* activity against *Aspergillus* species have

been developed and are routinely used for both prophylaxis and treatment in these patients.[4] Despite their widespread use and the increased awareness of this infectious complication by clinicians, the morbidity and mortality rate for immunocompromised patients with invasive aspergillosis remains unacceptably high. These clinical observations support the following statements:

- Persons with normal immune systems successfully keep *A. fumigatus* from developing invasive infections.
- Innate immunity is critical, as neutropenic patients have the highest risk of invasive aspergillosis.
- Successful recovery from invasive aspergillosis requires antimicrobial therapy coupled with some contribution from the host immune system.

These clinical observations lead to the fundamental question of what components of the immune response are critical to neutralize *A. fumigatus* quickly

doi: 10.1111/j.1749-6632.2012.06837.x

and efficiently. Knowledge of the basic pathways of antifungal defense in both normal and immuno-compromised animal models will likely elucidate the fundamental pathways necessary to develop novel therapeutics against this invasive fungal pathogen.

Innate immunity to *Aspergillus*

To provide a conceptual framework, the immune system can be divided into two principal branches: innate and adaptive immunity.[5] The innate immune response includes neutrophils, monocytes/macrophages, and dendritic cells (DCs) whose primary role is to engulf microorganisms before they cause harm.[6] To detect their presence, these phagocytic cells express both surface-disposed and endosomal Toll-like receptors (TLRs) that activate the cell in the presence of infectious agents. Instead of relying on highly specific signatures from pathogens TLRs sense the presence of common building blocks of microorganisms as triggers.[7] The outcome of TLR signaling is determined, in part, by cell lineage and the selective use of signaling adaptors.[8] Both ligand recognition by TLRs and the functional outcome of binding are governed, in part, by the subcellular location of the TLRs.[9]

In addition to TLRs, C-type lectin receptors (CLRs) compose a large family of receptors that bind to carbohydrates.[10] The lectin binding activity of these receptors is mediated by conserved carbohydrate-recognition domains, and members of this family of proteins participate directly in the host defense against fungal infections.[11] Dectin-1, a surface type II membrane protein and CLR, is highly expressed on phagocytes.[12] Dectin-1 recognizes the carbohydrate epitope β-1,3-glucan, which constitutes the major cell wall component of multiple pathogenic fungi, including *Candida albicans* and *A. fumigatus*.[13] Dectin-1 is required for proper modulation of immune responses against fungal pathogens, and patients with mutations in Dectin-1 are at higher risk for invasive fungal infections.[14,15] The cytoplasmic tail of Dectin-1 contains an immunoreceptor tyrosine-based activation (ITAM)-like motif.[16] Upon ligation of the extracellular domain of Dectin-1 the tyrosine residue within the cytoplasmic ITAM motif is phosphorylated. This phosphorylation event results in recruitment of spleen tyrosine kinase (Syk). Ultimately, this pathway leads to the activation of NF-κB, production of reactive oxygen species (ROS), and elaboration

of proinflammatory cytokines, including TNF-α and IL-12, which leads to recruitment of additional immune cells to the site of infection.[17] While the kinetics of this response is rapid (minutes to hours), the innate immune system does not provide memory to a given infection. In contrast, adaptive immunity consists of T and B lymphocytes that become activated after initiation of innate immunity and focus on specific peptides derived from pathogen-encoded proteins.[18] In order to extract pathogen-specific information, newly enveloped microorganisms are shuttled into membrane-delimited compartments termed *phagosomes*, delivered through the endocytic pathway by modification of the membrane proteins and intraphagosomal environment (*phagosome maturation*), and targeted to lysosomes for degradation.[19] Class II major histocompatibility complex (MHC) molecules intersect with these pathogen-containing compartments and relevant peptides derived from the pathogen are loaded into the MHC antigen binding clefts for delivery to the cell surface.[5,20] Much of our understanding of phagosomal biology has come from live cell imaging of professional antigen-presenting cells (APCs).[21]

Novel imaging tools to visualize host–pathogen interactions

Application of live-cell imaging tools to the interface of immunology and infectious disease is in its infancy though preliminary observations challenge our view of long-held beliefs about immune responses to microorganisms. As an example, to control *Mycobacterium tuberculosis*, the host forms granulomas, histologically distinct structures regularly seen in clinical tissue of infected persons. Conventional wisdom suggested that macrophages, T cells, and occasionally DCs form a static barrier that envelops organisms to control their spread. The use of intravital two-photon microscopy in the liver of mice infected with *M. bovis* demonstrated that immune cells showed remarkable dynamic movements to initiate and maintain these granulomas.[22] TNF-α–derived signals were required to recruit uninfected macrophages to these structures and preserve the involvement of T cells.

A long-term goal of our work is to understand how initial interactions between pathogens and phagocytic cells influence the degree of success of the host response. In order to understand this first step, it is important to visualize subcellular interactions of host innate immune cells with pathogenic organisms in order to observe phagocytosis. Successful phagocytosis not only neutralizes the threat from the pathogen, but also generates antigenic material crucial for the ensuing adaptive immune response. The regulation of phagocytosis and phagosome formation/maturation has significant implications in specific immune responses to pathogens.[23] To carry out these experiments, we have assembled both spinning-disk confocal microscopy and optical trapping on the same platform. The power of spinning disk confocal microscopy lies in the ability of the pinhole to exclude out-of-focal-plane fluorescence emission from biological specimens, allowing the high-contrast imaging of an optically sectioned slice. The scanning of confocal excitation light is achieved through two spinning disks in which one contains thousands of pinholes and the other contains an equal number of microlenses to focus the laser beam into the pinholes. This enables fast, multicolor, three-dimensional live cell imaging. In combination with submicron Z-plane scanning of the sample by use of a precise motorized stage, this system permits complete sectioning of the specimen to create a three-dimensional image.[20,24–27]

While this system generates illustrative images of APCs interacting with pathogens, spinning-disk confocal microscopes lack the capacity to exert control on the precise timing of these interactions. Commonly, investigators rely on serendipitous mixing of immune cells with pathogens to initiate phagocytosis. To overcome these obstacles, we adopted optical trapping to position actively live fungal pathogens at any arbitrary time and at any arbitrary location relative to immune cells, which allows real-time observation of the entire process using spinning disk confocal microscopy housed in an environment to mimic *in vivo* conditions.[28,29]

TLR9 is recruited to phagosomes containing *A. fumigatus*

The generation of fusion proteins of TLR/CLR appended to fluorescent proteins (including GFP and mCherry) coupled with live cell imaging has provided valuable insight into the dynamic nature of protein redistribution in phagosome formation/maturation.[30,31] While the majority of the data regarding TLR specificity and function have focused on bacterial and viral-derived ligands, our

understanding of fungal interactions with TLRs is accumulating.[1] Much of our understanding of TLR function is derived from infection experiments in mice lacking specific TLRs.[32] TLR2- and TLR4-defective mice show a decreased recruitment of neutrophils and a reduced cytokine response to *A. fumigatus* conidia and hyphae.[33] The complex fungal cell wall component zymosan triggers expression of proinflammatory cytokines via Dectin-1 and crosstalk with TLR2 and TLR4.[34] During swelling and germination of *A. fumigatus* conidia, β-1,3 glucans become exposed to the surface and can be targeted by cells of innate immune system expressing TLR2, TLR6, and Dectin-1, including macrophages and DCs.[35,36]

Comprehensive reviews of TLRs and the role of TLRs in fungal infections have been recently published.[1,6,8,18,37] We will focus on the role of TLR9 in the host defense of *A. fumigatus*. The subcellular localization of TLR9, which engages and signals to unmethylated CpG DNA, is tightly regulated and receptor activation is a multistep process.[38] TLR9 is translated into the endoplasmic reticulum in its mature, full-length form and then passes through the Golgi to the endolysosomal compartment where its ectodomain is proteolytically cleaved to generate a functional receptor.[39,40] In the endolysosomal compartment, ligand binding to preassembled TLR9 dimers induces a conformational change that allosterically initiates signal transduction.[41] While the truncated form of TLR9 can be found in the endolysosomal compartment of unstimulated cells, TLR9 trafficking has been shown to be a highly regulated, dynamic process.[31] The extent to which there is dynamic movement of TLR9 between subcompartments and the underlying processes regulating TLR9 trafficking remain poorly understood.

Although the best-known ligand for TLR9 is unmethylated bacterial and viral CpG-rich DNA, TLR9 has also been implicated in antifungal defense.[42,43] A role for TLR9 in host defense against *A. fumigatus* has been suggested by experiments in murine models of invasive pulmonary aspergillosis and allergic bronchopulmonary aspergillosis, where TLR9 modulates the innate immune response in the lung.[44] Although most studies have focused on the importance of TLR2 and TLR4 in defense against *A. fumigatus*, a polymorphism study associated increased susceptibility to ABPA with a polymorphism in the TLR9 gene.[45] Intranasal CpG, a known TLR9

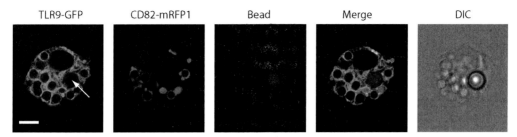

Figure 1. TLR9 is specifically recruited to phagosomes containing *A. fumigatus*. Mouse macrophages expressing TLR9-GFP (green) and CD82-mRFP1 (red) were exposed to both live *A. fumigatus* (Af293) and fluorescent polystyrene beads (blue 5 μm). Twenty minutes after phagocytosis, macrophages had taken up both particles, but only phagosomes containing *A. fumigatus* recruited both CD82-mRFP1 and TLR9-GFP. Scale bar equals 5 μm, and arrow denotes internalized polystyrene bead.

ligand, had a therapeutic effect during established murine fungal asthma, implicating a requisite role for TLR9 in this disease process.[46] However, the cell biological processes underlying TLR9-mediated *A. fumigatus* immune responses are still largely unresolved. TLR9-mediated recognition of *A. fumigatus* DNA by human and murine cells induced proinflammatory cytokines,[47] but the intracellular processes that enable *A. fumigatus* antigen recognition by TLR9 in professional APCs remain uninvestigated.

To gain insight into the intracellular fate of TLR9 when immune cells are exposed to *A. fumigatus* conidia, we investigated the spatiotemporal regulation of TLR9 compartmentalization after phagocytosis of *A. fumigatus* conidia by murine macrophages. We found that the presence of *A. fumigates* phagosomes resulted in a dramatic change of the subcellular distribution of TLR9 to a bright, ring-shaped compartment around the *A. fumigates* conidia (Fig. 1).[25] TLR9 recruitment was specifically induced by *A. fumigates* but not by bead-containing phagosomes, indicating that the composition of the phagocytosed content dictates recruitment of TLR9 to the phagosomal membrane. We demonstrated that *A. fumigatus*-induced TLR9 recruitment was independent of *A. fumigates* spore germination stage. Expression of TLR2, TLR4, and the TLR signaling adaptors MyD88 and TRIF were not required for successful *A. fumigatus* phagosomal TLR9 recruitment. Further investigation of the requirements for proper intracellular trafficking of TLR9 revealed that the TLR9 N-terminal proteolytic cleavage domain was critical for accumulation of TLR9 in CpG-containing compartments and *A. fumigates* phagosomal membranes.

In addition to *A. fumigatus*, we have demonstrated that specific triggering of TLR9 recruitment to the macrophage phagosomal membrane is a conserved feature of fungi of distinct phylogenetic origins, including *C. albicans, Saccharomyces cerevisiae, Malassezia furfur,* and *Cryptococcus neoformans.*[27] The capacity to trigger phagosomal TLR9 recruitment was not affected by a loss of fungal viability or cell wall integrity. TLR9 deficiency has been linked to increased resistance to murine candidiasis and to restriction of fungal growth *in vivo.* Macrophages lacking TLR9 demonstrate a comparable capacity for phagocytosis and normal phagosomal maturation compared to wild-type macrophages. We have shown that TLR9 deficiency increases macrophage TNF-α production in response to *C. albicans* and *S. cerevisiae,* independent of yeast viability.[27] The increase in TNF-α production was reversible by functional complementation of the TLR9 gene, confirming that TLR9 was responsible for negative modulation of the cytokine response. Consistently, TLR9 deficiency enhanced the macrophage effector response by increasing macrophage nitric oxide production. Moreover, microbicidal activity against *C. albicans* and *S. cerevisiae* was more efficient in TLR9 knockout macrophages than in wild-type macrophages.

In conclusion, our data have demonstrated that TLR9 is compartmentalized selectively to fungal phagosomes and negatively modulates macrophage antifungal effector functions. Our data support a model in which orchestration of antifungal innate immunity involves a complex interplay of fungal ligand combinations, host cell machinery rearrangements, and TLR cooperation and antagonism.

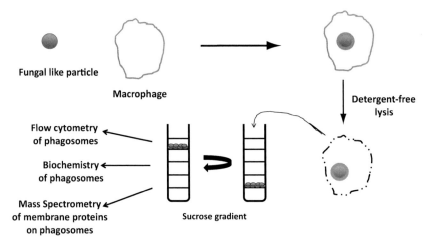

Figure 2. Cartoon depicting method to isolate and analyze phagosomes containing fungal-like particles. Polystyrene beads with fungal-derived carbohydrates covalently attached to the surface are incubated with macrophages. After a defined period, detergent-free lysis releases cellular content including intact phagosomes. Lysates are added to discontinuous sucrose gradients and then are subjected to ultracentrifugation. Phagosomes are easily isolated as a result of the buoyant properties of polystyrene beads and then analyzed by flow cytometry, Western blot, or mass spectrometry of membrane proteins.

Reducing the complexity: development and use of fungal-like particles

The mammalian immune response to invasive fungal pathogens, including *A. fumigates*, is a complex and dynamic process that serves to protect the host from invasive infection. Indeed, the generation of phagosomes and maturation of these organelles in part dictates the ensuing immune response to the pathogen. Our ability to understand the rules that govern this response is limited by two principal factors. First, the cell wall of *A. fumigatus* is a dynamic structure that undergoes both constitutional and morphologic changes, triggering additional responses from the APCs. *A. fumigatus* can swell inside the phagosome and reveal new antigenic determinants that are effectively shielded during the engulfment phase. Second, the resulting immune response from a complex particle such as *A. fumigatus* is difficult to dissect and to attribute specific responses to discrete cell wall components.

To overcome these obstacles, we set out to develop defined synthetic particles that displayed a single or fixed combination of fungal cell wall constituents on a size-matched polystyrene sphere.[48] In this way, we preserve the geometry of the native pathogen, an important feature to phagocytosis.[49] Synthetic fungal-like particles have multiple advantages as a substrate to probe the immune system. First, these particles do not vary during the phago-

somal maturation process. Second, the inflammatory response triggered by these particles can now be attributed to specific fungal ligands. Third, intrinsic properties of polystyrene beads permit the rapid recovery of these particles from APCs permitting direct interrogation of these phagosomes to probe the mammalian contribution of membrane proteins.[50] Polystyrene beads have a low buoyant density such that they float in concentrated sucrose solutions, allowing for the efficient separation of phagosomes from other cellular structures after cell lysis. Analysis of these phagosomes can be done by the use of flow cytometry (PhagoFACS),[51] Western blot analysis to determine the presence and relative quantity of a specific protein on the phagosome, and proteomic-based analysis to determine the composition of the mammalian phagosome generated by a synthetic ligand (Fig. 2).

As proof of principle, we have generated fungal-like particles that display β-1,3 glucan on the cell surface.[48] These fungal-like particles were characterized using differential interference contrast microscopy, immunofluorescence, and transmission-electron microscopy. The covalent interaction of the β-1,3-glucan layer to the surface of the bead was confirmed by a series of increasingly stringent detergent treatments. Purity of the β-1,3-glucan layer was also determined by incubating the beads with laminarinase, a specific β-1,3-gluconase.

By stimulating bone marrow–derived macrophages with conjugated β-1,3-glucan beads, we observed a dose-dependent response of TNF-α produced when compared to soluble β-1,3-glucan, uncoated beads, and soluble β-1,3-glucan mixed with uncoated beads. Finally, β-glucan coated fungal-like particles triggered dynamic and specific recruitment of GFP-Dectin-1 to nascent phagosomes in living mouse macrophages, demonstrating that the beads retained the desired biological properties.[48]

By providing a platform to probe directly immune responses to β-1,3-glucan, we can dissect the critical steps in the early recognition of pathogen-derived fungal carbohydrate antigens in innate immunity. Additionally, this system permits the addition of other relevant pathogen-derived carbohydrates to polymeric beads to analyze the specific contribution of these ligands in the immune response.

Concluding remarks

Three clinical outcomes from inhalation of *A. fumigates* exist: rapid removal of conidia without overt clinical symptoms (neutralization), exuberant Th2-biased responses characterized as allergic responses (allergic), or local tissue destruction that can lead to systemic infection (invasion). The outcome of this host–pathogen interaction appears to be dictated largely by host factors. Our understanding of the rules that govern the innate immune response to *A. fumigatus* conidia is relatively simple at this time, but this area is critical to the elucidation of the critical immune pathways that require modulation in order to add to our armamentarium against this clinically relevant pathogen. Novel tools utilizing subcellular imaging and synthetic fungal-like particles are critical developments to elucidate the molecular basis of *A. fumigatus*–host cell interactions.

Acknowledgments

J.M.V. and J.M.T. are supported by NIH Grant R01 AI092084. M.K.M. is supported by NIH Grant T32AI007061. The authors thank all of the current and former members of the laboratory.

Conflicts of interest

The authors declare no conflicts of interest.

References

1. Brown, G.D. & M.G. Netea. 2012. Exciting developments in the immunology of fungal infections. *Cell Host Microbe* **11**: 422–424.

2. Perfect, J.R. 2012. The impact of the host on fungal infections. *Am. J. Med.* **125**: S39–S51.

3. Pappas, P.G. *et al.* 2012. Invasive fungal infections among organ transplant recipients: results of the Transplant-Associated Infection Surveillance Network (TRANSNET). *Clin. Infect. Dis.* **50**: 1101–1111.

4. Herbrecht, R. *et al.* 2002. Voriconazole versus amphotericin B for primary therapy of invasive aspergillosis. *N. Engl. J. Med.* **347**: 408–415.

5. Vyas, J.M., A.G. Van der Veen & H.L. Ploegh. 2008. The known unknowns of antigen processing and presentation. *Nat. Rev. Immunol.* **8**: 607–618.

6. Roy, R.M. & B.S. Klein. 2012. Dendritic cells in antifungal immunity and vaccine design. *Cell Host Microbe* **11**: 436–446.

7. Casanova, J.L., L. Abel & L. Quintana-Murci. 2011. Human TLRs and IL-1Rs in host defense: natural insights from evolutionary, epidemiological, and clinical genetics. *Annu. Rev. Immunol.* **29**: 447–491.

8. Lee, C.C., A.M. Avalos & H.L. Ploegh. 2012. Accessory molecules for Toll-like receptors and their function. *Nat. Rev. Immunol.* **12**: 168–179.

9. Barton, G.M. & J.C. Kagan. 2009. A cell biological view of Toll-like receptor function: regulation through compartmentalization. *Nat. Rev. Immunol.* **9**: 535–542.

10. Sancho, D. & C. Reis e Sousa. 2012. Signaling by myeloid C-type lectin receptors in immunity and homeostasis. *Annu. Rev. Immunol.* **30**: 491–529.

11. van den Berg, L.M., S.I. Gringhuis & T.B. Geijtenbeek. 2012. An evolutionary perspective on C-type lectins in infection and immunity. *Ann. N.Y. Acad. Sci.* **1253**: 149–158.

12. Kerrigan, A.M. & G.D. Brown. 2011. Syk-coupled C-type lectins in immunity. *Trends Immunol.* **32**: 151–156.

13. Netea, M.G. & L. Marodi. 2010. Innate immune mechanisms for recognition and uptake of Candida species. *Trends Immunol.* **31**: 346–353.

14. Cunha, C. *et al.* 2010. Dectin-1 Y238X polymorphism associates with susceptibility to invasive aspergillosis in hematopoietic transplantation through impairment of both recipient- and donor-dependent mechanisms of antifungal immunity. *Blood* **116**: 5394–5402.

15. Ferwerda, B. *et al.* 2009. Human dectin-1 deficiency and mucocutaneous fungal infections. *N. Engl. J. Med.* **361**: 1760–1767.

16. Ariizumi, K. *et al.* 2000. Identification of a novel, dendritic cell-associated molecule, dectin-1, by subtractive cDNA cloning. *J. Biol. Chem.* **275**: 20157–20167.

17. Drummond, R.A. *et al.* 2011. The role of Syk/CARD9 coupled C-type lectins in antifungal immunity. *Eur. J. Immunol.* **41**: 276–281.

18. Casadevall, A. & L.A. Pirofski. 2012. Immunoglobulins in defense, pathogenesis, and therapy of fungal diseases. *Cell Host Microbe* **11**: 447–456.

19. Kagan, J.C. & A. Iwasaki. 2012. Phagosome as the organelle linking innate and adaptive immunity. *Traffic* **13**: 1053–1061.

20. Vyas, J.M. *et al.* 2007. Tubulation of class II MHC compartments is microtubule dependent and involves multiple endolysosomal membrane proteins in primary dendritic cells. *J. Immunol.* **178**: 7199–7210.

21. Bajenoff, M. & R.N. Germain. 2007. Seeing is believing: a focus on the contribution of microscopic imaging to our

understanding of immune system function. *Eur. J. Immunol.* **37**(Suppl 1): S18–S33.

22. Egen, J.G. *et al.* 2008. Macrophage and T cell dynamics during the development and disintegration of mycobacterial granulomas. *Immunity* **28**: 271–284.

23. Nusse, O. 2011. Biochemistry of the phagosome: the challenge to study a transient organelle. *ScientificWorld J* **11**: 2364–2381.

24. Artavanis-Tsakonas, K. *et al.* 2006. Recruitment of CD63 to Cryptococcus neoformans phagosomes requires acidification. *Proc. Natl. Acad. Sci. USA* **103**: 15945–15950.

25. Kasperkovitz, P.V., M.L. Cardenas & J.M. Vyas. 2010. TLR9 is actively recruited to Aspergillus fumigatus phagosomes and requires the N-terminal proteolytic cleavage domain for proper intracellular trafficking. *J. Immunol.* **185**: 7614–7622.

26. Artavanis-Tsakonas, K. *et al.* 2011. The tetraspanin CD82 is specifically recruited to fungal and bacterial phagosomes prior to acidification. *Infect. Immun.* **79**: 1098–1106.

27. Kasperkovitz, P.V. *et al.* 2011. Toll-like receptor 9 modulates macrophage antifungal effector function during innate recognition of Candida albicans and Saccharomyces cerevisiae. *Infect. Immun.* **79**: 4858–4867.

28. Tam, J.M. *et al.* 2010. Control and manipulation of pathogens with an optical trap for live cell imaging of intercellular interactions. *PLoS One* **5**: e15215.

29. Tam, J.M. *et al.* 2011. Use of an optical trap for study of host-pathogen interactions for dynamic live cell imaging. *J. Vis. Exp.* **53**: e3123, doi: 10.3791/3123.

30. Underhill, D.M. *et al.* 2005. Dectin-1 activates Syk tyrosine kinase in a dynamic subset of macrophages for reactive oxygen production. *Blood* **106**: 2543–2550.

31. Kim, Y.M. *et al.* 2008. UNC93B1 delivers nucleotide-sensing toll-like receptors to endolysosomes. *Nature* **452**: 234–238.

32. Carpenter, S. & L.A. O'Neill. 2007. How important are Toll-like receptors for antimicrobial responses? *Cell Microbiol.* **9**: 1891–1901.

33. Meier, A. *et al.* 2003. Toll-like receptor (TLR) 2 and TLR4 are essential for Aspergillus-induced activation of murine macrophages. *Cell Microbiol.* **5**: 561–570.

34. Ferwerda, G. *et al.* 2008. Dectin-1 synergizes with TLR2 and TLR4 for cytokine production in human primary monocytes and macrophages. *Cell Microbiol.* **10**: 2058–2066.

35. Faro-Trindade, I. *et al.* 2012. Characterisation of innate fungal recognition in the lung. *PLoS One* **7**: e35675.

36. Hohl, T.M. *et al.* 2005. Aspergillus fumigatus triggers inflammatory responses by stage-specific beta-glucan display. *PLoS Pathog.* **1**: e30.

37. Li, X., S. Jiang & R.I. Tapping. 2010. Toll-like receptor signaling in cell proliferation and survival. *Cytokine* **49**: 1–9.

38. Wagner, H. 2004. The immunobiology of the TLR9 subfamily. *Trends Immunol.* **25**: 381–386.

39. Ewald, S.E. *et al.* 2008. The ectodomain of Toll-like receptor 9 is cleaved to generate a functional receptor. *Nature* **456**: 658–662.

40. Park, B. *et al.* 2008. Proteolytic cleavage in an endolysosomal compartment is required for activation of Toll-like receptor 9. *Nat. Immunol.* **9**: 1407–1414.

41. Latz, E. *et al.* 2007. Ligand-induced conformational changes allosterically activate Toll-like receptor 9. *Nat. Immunol.* **8**: 772–779.

42. Bellocchio, S. *et al.* 2004. TLRs govern neutrophil activity in aspergillosis. *J. Immunol.* **173**: 7406–7415.

43. Miyazato, A. *et al.* 2009. Toll-like receptor 9-dependent activation of myeloid dendritic cells by Deoxynucleic acids from Candida albicans. *Infect. Immun.* **77**: 3056–3064.

44. Ramaprakash, H. *et al.* 2009. Toll-like receptor 9 modulates immune responses to Aspergillus fumigatus conidia in immunodeficient and allergic mice. *Infect. Immun.* **77**: 108–119.

45. Carvalho, A. *et al.* 2008. Polymorphisms in toll-like receptor genes and susceptibility to pulmonary aspergillosis. *J. Infect. Dis.* **197**: 618–621.

46. Ramaprakash, H. & C.M. Hogaboam. 2010. Intranasal CpG therapy attenuated experimental fungal asthma in a TLR9-dependent and -independent manner. *Int. Arch. Allergy Immunol.* **152**: 98–112.

47. Ramirez-Ortiz, Z.G. *et al.* 2008. Toll-like receptor 9-dependent immune activation by unmethylated CpG motifs in Aspergillus fumigatus DNA. *Infect. Immun.* **76**: 2123–2129.

48. Tam, J.M. *et al.* 2012. Use of fungal derived polysaccharide-conjugated particles to probe Dectin-1 responses in innate immunity. *Integr. Biol.* **4**: 220–227.

49. Champion, J.A. & S. Mitragotri. 2006. Role of target geometry in phagocytosis. *Proc. Natl. Acad. Sci. USA* **103**: 4930–4934.

50. Stuart, L.M. *et al.* 2007. A systems biology analysis of the Drosophila phagosome. *Nature* **445**: 95–101.

51. Cebrian, I. *et al.* 2011. Sec22b regulates phagosomal maturation and antigen crosspresentation by dendritic cells. *Cell* **147**: 1355–1368.